Acknowledgement

This symposium was sponsored
by the Japan Society for the Promotion of Science
and this volume was edited for
the Restoration Medical Engineering Program.

Tissue Engineering for Therapeutic Use 1

Tissue Engineering for Therapeutic Use 1

Proceedings of the First International Symposium of Tissue Engineering for Therapeutic Use, Kyoto, Japan, 2 March 1997

Editor:

Yoshito Ikada
Research Center for Biomedical Engineering
Kyoto University, Kyoto, Japan

Co-Editor:

Yoshio Yamaoka
Department of Gastroenterological Surgery
Kyoto University Graduate School of Medicine, Kyoto, Japan

1998

ELSEVIER

Amsterdam – Lausanne – New York – Oxford – Shannon – Singapore – Tokyo

ELSEVIER SCIENCE B.V.
Sara Burgerhartstraat 25
P.O. Box 211, 1000 AE Amsterdam, The Netherlands

First edition 1998

Library of Congress Cataloging in Publication Data
A catalog record from the Library of Congress has been applied for.

International Congress Series No. 1169
ISBN: 0 444 82993 8

∞ The paper used in this publication meets the requirements of ANSI/NISO Z39.48-1992 (Permanence of Paper).

Printed in the Netherlands

Preface: the restoration medical engineering in Japan

The Japan Society for the Promotion of Science (JSPS) has launched a new, far-sighted program entitled "Research for the Future (RFTF)". This program, which started in FY 1996, is funded through investment made by the Japanese government to promote and expand the frontiers of scientific research. The Research for the Future Program can be characterized by the following objectives:

1. To promote highly creative, future-oriented research that will produce the intellectual assets and form the scientific bases for advancing the development of various sectors in the years leading up to the 21st century.
2. To carry out research aimed at meeting society's diversifying needs at universities and other research institutions, with cooperation from the industrial and other sectors when expedient.
3. To foster young postdoctoral researchers by employing them in RFTF projects as "JSPS Research Associates".
4. To advance research through formats that combine outstanding intellectual resources with international perspectives, and that promote international collaboration.

Based on the aforementioned objectives of the RFTF program, JSPS's RFTF Program Committee designates the specific research fields to be pursued and sets up Research Promotion Committees to advance the research activities in each of the fields.

In order to study and make decisions on important policy issues relevant to the program, the aforementioned RFTF Program Committee has been established within JSPS. It comprises mainly senior leaders of academia. Under the Program Committee, there are three subcommittees that have been set up to deliberate issues related specifically to the program's three academic domains: these are the Physical and Engineering Sciences Subcommittee, the Life Sciences Subcommittee, and the Integrated Fields Subcommittee. The Research Promotion Committees are organized by the Program Committee and consist of five to 10 leading scientists in the respective research fields. In FY 1996, 17 Research Promotion Committees were organized.

The "Integrated Fields" includes three fields: information on life systems, restoration medical engineering, and production on novel useful materials through integration of life sciences and chemical techniques. The "Restoration Medical Engineering" covers the five projects shown in Table 1 overleaf.

Yoshito Ikada
Chairman, Kyoto University

Table 1.

Research project	Project leader
Biomaterials for tissue engineering	Teruo Okano (Tokyo Women's Medical College)
Bioprocess engineering of functional regeneration of cultured cells	Norio Oshima (University of Tsukuba)
Tissue engineering for soft tissues	Yasuhiko Shimizu (Kyoto University)
Tissue engineering for organs	Yoshio Yamaoka (Kyoto University)
Tissue engineering for hard tissues	Shoji Enomoto (Tokyo Medical and Dental University)

Tissue Engineering for Therapeutic Use 1.
Y. Ikada and Y. Yamaoka, editors.

Biomaterials for tissue engineering

Teruo Okano
Institute of Biomedical Engineering, Tokyo Women's Medical collage, Tokyo, Japan

Clearly, the role of extracellular matrix in tissue formation and differentiation is complex, but extremely important to be understood. As increasingly more has become known about these roles, interfacial interactions of biomaterials with cells has attracted much attention from a diverse spectrum of biomedical fields. In particular, development of new therapeutic strategies is one important application area for biomaterials research. The aim of our research is to establish new concepts for novel, effective therapies by controlling cell functions through contact-induced interactions between cells and biomaterials' surfaces.

The understanding and control of nonspecific interactions between cells and polymeric materials is an important subject for tissue engineering. To control nonspecific interactions, we have studied substrate molecular architectural effects on cell-biomaterial interactions. Interactions of synthetic polymer surfaces with proteins, cells and tissue have been studied systematically using polymers with well-defined structures. Sizes and morphologies of hydrophilic/hydrophobic microdomain structures have been regulated through the control of both composition and block sequences in synthetic block copolymers. Surface microdomain structures were found to show interesting characteristics that inhibit surface-induced platelet adhesion and aggregation. In particular, block copolymers constructed from 2-hydroxyethyl methacrylate (HEMA) and styrene (St) showed excellent nonthrombogenicity both in vitro [1–3] and in vivo [4,5]. Small-diameter vascular grafts coated with HEMA-St block copolymer films maintain monolayer protein adsorption on the interior lumen exposed to blood, successfully stabilizing blood-polymer interfaces for longer than 1 year. From these results, a role for polymer surface micro-architecture in biointerfacial design was proposed and developed. The author has proposed a surface-mediated "microdomain effect" to achieve ideal nonthrombogenic surfaces. Hydrophilic polymer-grafted surfaces have also been shown to resist cell adhesion by minimizing nonspecific interactions of cells with polymer surfaces. Based on these concepts, adhesion of platelets and other cell types using our two surface archetypes will be studied to minimize nonspecific interactions.

Introduction of stimuli-responsive synthetic polymers as switching sequences into both artificial materials and bioactive molecules permits external stimuli-induced modulation of their structures and "on-off" switching of their respective functions at molecular levels. Our research was directed toward the fundamental studies and development of biomedically relevant modulation systems using the temperature-responsive polymer, poly(N-isopropylacrylamide) (PIPAAm) bearing carboxyl or amino end groups, as switching sequences [6]. This polymer was attached to solid surfaces, or to cross-linked hydrogels to create new PIPAAm-grafted surfaces [7] and graft type hydrogels [8,9], respectively. We have reported significant hydrophilic/hydrophobic surface property alterations in response to external temperature changes using PIPAAm-grafted solid surfaces. Specifically, these surfaces demonstrate hydrophobic properties at temperatures above PIPAAm's lower critical solution temperature (LCST). Above the LCST, PIPAAm chains exist in collapsed conformations. Grafted chain architectures affect temperature-responsive surface functional changes and their kinetics of change. Single-end attachment of PIPAAm chains to solid surfaces, leaving the other end mobile, results in surfaces showing large and rapid thermally modulated hydrophilic/hydrophobic changes. Activation and inactivation of platelets are likewise controlled by temperature: platelets show no surface interaction below the grafted surface LCST and are readily adhered to the PIPAAm-grafted surfaces above the LCST. Hepatocytes and endothelial cells readily attach and proliferate on PIPAAm-grafted surfaces at 37°C (above LCST). When the temperature is lowered below the LCST, surface-grafted PIPAAm molecules hydrate and the surface becomes hydrophilic. Cultured cells readily and spontaneously detach from the hydrated surfaces at this lower temperature. As a result, cultured cells are easily recovered from the PIPAAm-grafted surfaces by low-temperature treatment. Cells harvested by this method exhibit improved cell viability and functionality compared to those harvested by traditional enzymatic treatments [10].

To establish two-dimensional manipulation for confluent cultures of endothelial cells and/or hepatocytes, PIPAAm-grafted surfaces with higher temperature-sensitivity are studied in terms of surface composition and grafted polymer structure. Confluent cultures of hepatocytes, endothelial cells and other epithelial cells were recovered from PIPAAm-grafted culture dishes as cell monolayer sheets by low-temperature treatment, as shown in Fig. 1. Two different methods were developed to transfer cell sheets. Chitin membranes were developed to transfer cell sheets during the transfer procedure. First, cell sheets are adsorbed onto the chitin membranes, then the supported cells are transferred to a new substrate, leaving the apical surface exposed to the culture medium. The other method is the direct contact of the apical side of the cell sheet with new surface, and as a result, the original apical side faces the new basal side after transfer. Based on this cell sheet recovery technique using PIPAAm-grafted culture substrates, double-layer, triple-layer, and multilayers of cell sheets can be prepared with or without the use of cell adhesive proteins between cell sheets. Effects of growth factors and other cytokines on the function of multilayered cell sheets will also be investigated. In particular, the possibility of co-culture using combinations of two different types of cell sheets will be studied. Through these studies, transplantable epithelial cell sheets and keratinocytes will be developed. Tissues of pancreatic B islet cells as well as hepatocytes maintain a healthy, high metabolic function when harvested from PIPAAm-grafted surfaces and transplanted in

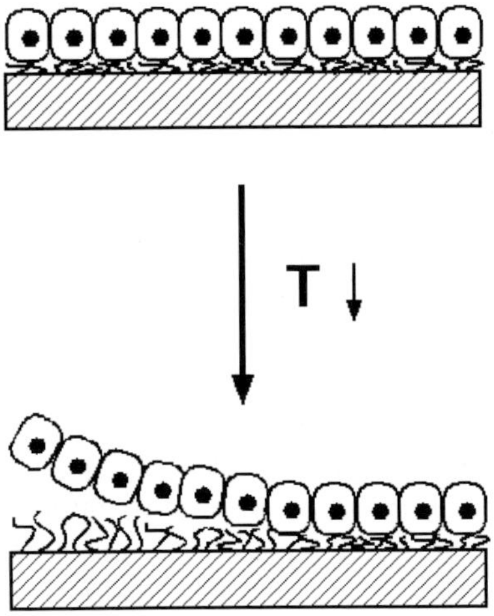

Fig. 1. Cell recovery by temperature-responsive culture dishes. Cell sheets of hepatocytes or endothelial cells are easily recovered by temperature-responsive culture dishes.

vivo. Microglia are also being investigated for recovery from PIPAAm-grafted culture surfaces while maintaining their structures and specific functions. From these experiments, general concepts and new techniques to manipulate cells and cell sheets are being established as new tools for tissue engineering.

Phenylboronic acid is known to form reversible complexes with polyol compounds where cis hydroxyl groups are in the proper proximal distance. Our group has already established formation of specific complexes of phenylboronic acid copolymers with sugar moieties under physiological conditions [11,12]. Interaction of various cell types with phenylboronic acid-containing polymers are currently being investigated to control cell functions through specific interactions. Focusing on the lectin-mimicking function of phenylboronic acid in terms of binding to sugar moieties on cell membranes, phenylboronic acid-induced T lymphocyte activation is studied. Phenylboronic acid copolymers also induce spontaneous capillary formation of a confluent cultured endothelial cell. In this project, effects of phenylboronic acid will be investigated as an artificial biological response modifier to achieve an improved immune response in vivo and tissue formation of endothelial cells and hepatocytes in vitro.

References

1. Okano T, Nishiyama S, Shinohara I, Akaike T, Sakurai Y, Kataoka K, Tsuruta T. Effect of hydrophilic and hydrophobic microdomains on mode of interaction between block polymer and blood platelets. J Biomed Mater Res 1981;15(3):393–402.
2. Okano T, Aoyagi T, Kataoka K, Abe K, Sakurai Y, Shimada M, Shinohara I. Hydrophilic-hydrophobic microdomain surfaces having an ability to suppress platelet aggregation and their in vitro antithrombogenicity. J Biomed Mater Res 1986;20(7):919–927.
3. Okano T, Uruno M, Sugiyama N, Shimada M, Shinohara I, Kataoka K, Sakurai Y. Suppression

of platelet activity on microdomain surfaces of 2-hydroxyethyl methacrylate-polyether block copolymers. J Biomed Mater Res 1986;20(7):1035–1047.

4. Yoshida R, Sakai K, Okano T, Sakurai Y. Characteristics of drug release using heterogeneous polymer. Drug Deliv Syst 1990;5(4):273–277.
5. Nojiri C, Okano T, Takemura N, Senshu K, Kido T, Koyanagi H, Kim SW, Akutsu T. Improved patency of HEMA/styrene block copolymer-coated small vessel prosthesis without neointima formation. In: Akutsu T, Koyanagi H (eds) Artificial Heart 4, Heart Replacement; 1993;53–60.
6. Okano T, Yui N, Yokoyama M, Yoshida R. Advances in polymeric systems for drug delivery. Jpn Technol Rev, Section E: Biotechnol 1994;4(2) (Gordon and Breach Science Publishers).
7. Takei YG, Aoki T, Sanui K, Ogata N, Sakurai Y, Okano T. Temperature-modulated platelet and lymphocyte interactions with poly(N-isopropylacrylamide)-grafted surfaces. Biomaterials 1995; 16:667–673.
8. Yoshida R, Uchida K, Kaneko Y, Sakai K, Kikuchi A, Sakurai Y, Okano T. Comb-type grafted hydrogels with rapid de-swelling response to temperature changes. Nature 1995;374(6519): 240–242.
9. Kaneko Y, Sakai K, Kikuchi A, Yoshida R, Sakurai Y, Okano T. Influence of freely mobile grafted chain length on dynamic properties of comb-type grafted poly(N-isopropylacrylamide) hydrogels. Macromolecules 1995;28(23):7717–7723.
10. Okano T, Yamada N, Sakai H, Sakurai Y. A novel recovery system for cultured cells using plasma-treated polystyrene dishes grafted with poly(N-isopropylacrylamide). J Biomed Mater Res 1993; 27(10):1243–1251.
11. Miyazaki H, Kikuchi A, Koyama Y, Okano T, Sakurai Y, Kataoka K. Boronate-containing polymer as novel mitogen for lymphocytes. Biochem Biophys Res Commun 1993;195(2):829–836.
12. Aoki T, Nagao Y, Terada E, Sanui K, Ogata N, Yamada N, Sakurai Y, Kataoka K, Okano T. Endothelial cell differentiation into capillary structures by copolymer surfaces with phenyboronic acid groups. J Biomater Sci Polym Edt 1995;7(7):539–550.

Tissue Engineering for Therapeutic Use 1.
Y. Ikada and Y. Yamaoka, editors.

Bioprocess engineering of functional regeneration of cultured animal cells

Project Leader

- Norio Ohshima PhD, Professor of Biomedical Engineering, Institute of Basic Medical Sciences, University of Tsukuba, Tsukuba Science City, Ibaraki, Japan

A large number of hybrid-type artificial organs that utilize cultured animal cells to replace functions of deteriorated organs and tissues have been tried, nevertheless, most of them have been proven to have limited applicability for therapeutic use. These insufficient results in developing such types of bioartificial organs are due to the lack of engineering viewpoints, especially in the fields of chemical engineering, bioprocess engineering and tissue engineering. To develop clinically applicable bioartificial organs, scaled-up bioreactor systems utilizing cultured cells are indispensable. Moreover, most recent development of fundamental cellular engineering approaches, basic medical sciences and biomaterial sciences also require incorporation.

This project aims to promote basic and applied research toward an ultimate goal for the development of bioartificial organs utilizing cultured animal cells. As a target of research objects, individual research groups are concerned with either hepatocytes, chondrocytes and bone marrow cells, or vascular endothelial cells under three major subthemes as described below.

Functional regeneration of cultured hepatocytes

Members

- Kennichi Yanagi (Dept. Biomedical Engineering, Univ. Tsukuba)
- Hirotoshi Miyoshi (Dept. Biomedical Engineering, Univ. Tsukuba)
- Katashi Fukao (Dept. Surgery, Univ. Tsukuba)
- Hideki Taniguchi (Dept. Surgery, Univ. Tsukuba)
- Shinichi Kasai (Dept. Surgery, Asahikawa Medical College)
- Pei Kan (Dept. Biomedical Engineering, Univ. Tsukuba)

Address for correspondence: Norio Ohshima PhD, Professor of Biomedical Engineering, Institute of Basic Medical Sciences, University of Tsukuba, Tsukuba Science City, Ibaraki 305, Japan. E-mail: ohshima@md.tsukuba.ac.jp

To develop an artificial liver support system, effective culture techniques that can attain a high density culture of hepatocytes and that can maintain metabolic functions specific to the liver in vitro are to be investigated. The use of highly porous material as a culture substratum has already been proven to be promising in constructing a clinically feasible bioartificial liver, thus, scaling-up of such a type of bioreactor module, and modification of the culture method are to be investigated. For clinical use, methods of cell harvesting used in order to obtain a large quantity of porcine hepatocytes and their cryopreservation are also aimed at being established.

A fundamental approach to support artificial organ development is also pursued. To identify hepatic stem cells that can proliferate and differentiate to the hepatocytes, an in vivo assay system is aimed at being established by hepatocyte transplantation experiments using β-galactosidase-gene-engineered transgenic mice.

Functional regeneration of chondrocytes and bone marrow cells

Members

- Hiromitsu Nakauchi (Dept. Immunology, Univ. Tsukuba)
- Masayuki Yamamoto (Dept. Molecular and Developmental Biology, Univ. Tsukuba)
- Tetsuya Tateishi (Bionic Design Research Group, National Institute for Advanced Interdisciplinary Research, AIST, MITI)
- Toshimasa Uemura (Bionic Design Research Group, National Institute for Advanced Interdisciplinary Research, AIST, MITI)

As fundamental research to establish effective means for regeneration of cartilage, effects of physical and biomechanical factors, such as hydrostatic pressure and oxygen gradient to the medium, on chondrocyte functions are examined.

Furthermore, development of an in vitro culture system of blood cells which can replace hematopoietic functions in the bone marrow will be undertaken, and its application to a new tissue regeneration therapy to treat intractable blood diseases will be attempted.

Functional regeneration of neovasculature and blood vessels

Members

- Keiko Ookawa-Chinzei (Dept. Biomedical Engineering, Univ. Tsukuba)
- Teruo Watanabe (Dept. Pathology, Univ. Tsukuba)
- Jiang-Lin Fan (Dept. Pathology, Univ. Tsukuba)
- Katsutoshi Goto (Dept. Pharmacology, Univ. Tsukuba)
- Kazuhiko Igarashi (Dept. Molecular & Developmental Biology, Univ. Tsukuba)

- Takashi Ushida (Bionic Design Research Group, National Institute for Advanced Interdisciplinary Research, AIST, MITI)
- Yoichiro Abe (Dept. Pharmacology, Univ. Tsukuba)

Mechanisms of angiogenesis or regeneration of injured blood vessels are to be investigated. The roles of vascular endothelial cells and smooth muscle cells in restoring vascular injury are elucidated through in vivo animal experiments by use of transgenic rabbits, and in vitro perfusion experiments with cultured vascular endothelial cells.

The pathophysiology of small blood vessels in the central nervous system, "blood-brain barrier", are examined with special emphasis on the roles of astrocytes in experimental brain edema in rats.

A mechanism of functional regeneration of organs and tissues in response to a reduced supply of oxygen is also investigated. Furthermore, the molecular mechanism of functional regeneration of vascular endothelial cells is elucidated from the viewpoint of functions of transcription factors, which regulate cell differentiation.

Tissue Engineering for Therapeutic Use 1.
Y. Ikada and Y. Yamaoka, editors.

Tissue engineering for soft tissues

Project Leader

- Yasuhiko Shimizu MD, Department of Artificial Organs, Research Center for Biomedical Engineering, Kyoto University

Core Members

- Tatsuo Nakamura MD, Department of Artificial Organs, Research Center for Biomedical Engineering, Kyoto University
- Naohide Tomita MD, Department of Artificial Organs, Research Center for Biomedical Engineering, Kyoto University
- Katsuaki Endo MD, Department of Brain Science, Medical Faculty, Kyoto University

Purpose of study

None of the tissues or organs of mammals including man are capable of regenerating their normal shape after total or partial removal by surgery or trauma. The only exception is the liver, which is well known to show marked regeneration. Therefore, almost all artificial organs used hitherto have been composed of synthetic nonabsorbable artificial materials to replace the failing organs. Conventional artificial kidneys, i.e., dialyzers, or artificial heart-lung machines are used only as temporary extracorporeal systems and are too large to be implanted into the patient's body. Moreover, they can cover only several of the natural functions of the original organs. Although there have been a few attempts to implant such organs into patients, the results have been far from satisfactory and they are now not clinically accepted because they are rejected due to foreign body reactions in the long term. Thus the concept of replacing human organs with artificial materials is fundamentally limited. On the other hand, although natural organ transplantation has many advantages (namely, the implanted organs are compact and retain all their functions), it has not yet been accepted in Japan. The main reasons are the absolute shortage of donors and the difficulty in accepting the concept of brain death, which is rooted deeply in our traditional way of thinking. Moreover, even if transplantation is carried out successfully, the recipients cannot avoid the troublesome side effects of immunosuppressive drugs during the

Address for correspondence: Yasuhiko Shimizu MD, Department of Artificial Organs, Research Center for Biomedical Engineering, Kyoto University, Kyoto, Japan. Tel.: +81-75-751-4149. Fax: +81-75-751-4844. E-mail: shiko@medeng.kyoto-u.ac.jp remainder of their lives. Thus, organ transplantation has not been universally clinically accepted.

In view of this situation, we have devised a completely new approach to the development of artificial organs. The main procedure using tissue engineering for soft tissues involves the removal of the cell component from auto- or allo-organs to obtain only the extracellular matrix, the so-called refined extracellular matrix (ECM), and reconstitutes the solid structure from the extracted collagen. This ECM or reconstituted structure is then employed as scaffolding, which after implantation into the patients is used for the regeneration or redifferentiation of tissue. Organs made of self-cells thus regenerate. Organs that regenerate in this manner not only possess highly differentiated tissue structures, but also show functional recovery, because all the cells are derived from the patients themselves. Whether or not our new method is practicable will depend mainly on the intrinsic regeneration capacity of each tissue. Up until now, in higher mammals, including man, it has been believed that highly differentiated organs lose their ability to regenerate. We consider that mammals do not, in fact, lose this potential, and that the potential is hidden by excessively rapid wound healing around the failing tissues. In this sense, if we can provide good conditions using refined ECM, we can induce this hidden potential even in higher mammals. We have already carried out successful trials at regenerating peripheral nerves, the esophagus, the trachea and blood vessels with this method. A similar method is also applicable to other soft tissue organs such as the liver, heart and lung, as well as the spinal cord. These results will be welcomed by patients who are dependent on palliative life-support systems, or transplantation candidates who are waiting for suitable donors. An additional benefit is that patients will be freed from the side effects of immunosuppressive drugs. The judgment of brain death can then be discussed separately from the issue of transplantation, and will become a personal problem. Furthermore, this new approach helps to reduce ever-expanding medical costs, which are in danger of destroying our health insurance system in the near future.

No study based on these concepts has ever been carried out either in Japan or abroad. In this sense, our pioneering work is expected to be a major area of medical science for the coming generation.

Outline of study plan

The target organs currently being considered for this development project are the heart, heart valves, esophagus, trachea, peripheral nerves, spinal cord, cornea, tendons and ligaments. We plan to employ the three methods as described below.

ECM method

To obtain the purified extracellular matrix, cell components are completely removed from homo- or allo-organs. The solid structure is reconstituted from the ECM and extracted collagen. Growth factors are then applied to facilitate cell proliferation. Then this ECM-collagen-growth factor composite is implanted into the living body as a temporary scaffolding for new organ regeneration. Additionally, bioabsorbable materials will also be applied instead of purified ECM as a bulk structure for organ regeneration. Both extracted collagen and growth factors should facilitate cell proliferation and cell redifferentiation, leading to regeneration of organs completely composed of cells derived from patients.

Cell + ECM method

Cells (or living tissues) of patients are complexed (mixed) with purified ECM or bioabsorbable material. Using this complex, reconstruction of the failing tissues or organs will be attempted.

Transgenic method

Human genes are inserted into fertilized animal ova to make transgenic animals. The corresponding genes are expressed as receptors on the cell surface which are involved in rejection. The organs or tissues of these transgenic animals are then harvested for use as bioartificial organs, which are almost immunologically inert, especially lacking the super acute rejection caused by complement. In xenotransplantation, such an organ is transplanted into humans directly. In our project, all the cell components are removed from the harvested organ, and then only the ECM from the transgenic animals is used as a scaffolding for the self-cell regeneration. Accordingly, the regenerated organ consists of autologous cells alone. This is the main difference from ordinary xenotransplantation.

Supported by the current rapid progress being made in biological science and tissue engineering, our challenge aimed at new artificial organs has just started. This is expected to be the main research theme of medical science in the 21st century, since it is based on a completely new concept, as mentioned above.

Tissue Engineering for Therapeutic Use 1.
Y. Ikada and Y. Yamaoka, editors.

Tissue engineering for organ regeneration

Project Leader

- Yoshio Yamaoka MD Professor, Department of Gastroenterological Surgery, Postgraduate School of Medicine, Kyoto University

Core Member

- Hiroo Iwata MD Associate Professor, Research Center for Biomedical Engineering, Kyoto University
- Kazutomo Inoue MD Associate Professor, Department of Oncological Surgery, Postgraduate School of Medicine, Kyoto University
- Iwao Ikai MD Assistant Professor, Department of Gastroenterological Surgery, Kyoto University

Collaborating Member

- Yoshihito Honda MD Professor, Department of Ophthalmology and Visual Science, Postgraduate School of Medicine, Kyoto University
- Kazuwa Nakao MD Professor, Department of Medicine and Clinical Science, Postgraduate School of Medicine, Kyoto University
- Koichi Tanaka MD Professor, Department of Transplantation Immunology, Postgraduate School of Medicine, Kyoto University
- Osamu Yoshida MD Professor, Department of Urology, Postgraduate School of Medicine, Kyoto University
- Shigeki Hitomi MD Professor, Department of Surgery, Chest Disease Research Institute, Kyoto University

Functional failure of a main organ introduces death to life. Insufficiency of the sensory organs critically disturbs the quality of life. Much research has been focussed on how to support the insufficient organs using biotechnological approaches. Partially successful in this theme is the mechanical support for respiratory insufficiency or the dialysis system for renal failure. Liver failure or pancreatic insufficiency could not be overcome by traditional bioengineering due to their complicated function. Organ transplantation solves this kind of probem, however, another big problem is the shortage of donor organs, and also the continuous medication for immunosuppression. Considering these factors, the purpose of this project is 1) to prepare an artificial organ using tissues or cells from human beings or other animal combinations with artificial materials including biomaterials, and 2) to use the artificial organ — rescue of the targeted organ would be expected only by regeneration of its own cells and refunction as the native organ.

Regeneration of retina

The process of retinal pigment epithelial healing after Bruch's membrane injury and experimental retinal injury was followed by tissue biochemical staining. Recently, retinal cell transplantation has been planned using the stem cells obtained from the hippocampus of the brain.

Regeneration of the cardiovascular system

The roll of adhesion molecules to the extracellular matrix of myoblast is under examination aiming to regenerate tile cardiac muscle. The next plan is for dedifferentiated injured vessels due to aging or adult diseases to be introduced to redifferentiation for neovascularization for which the functional recovery of the related area would be expected.

Regeneration of liver

Two clinically relevant liver assist devices were introduced using xenogenic tissue, extracorporeal liver xenoperfusion (ECLP) and a hybrid artificial liver. In ECLP, we procured porcine whole liver amounting to 450 g and perfused them with our original system. In the hybrid liver, 50 g isolated porcine hepatocytes were inoculated into the inner space of the follow fiber whose cutoff value is 32 kDa. Both modalities were perfused with human whole blood. In the perfusion experiments, the hybrid liver showed a long-term detoxication capacity for ammonia, lidocaine, and galactose for more than 2 weeks. Inoculated hepatocytes formed desmosome within 12 h. With our ECLP system, liver xenoperfusion could be continued for longer than 6 h without evident humoral injury. The xenoperfusion liver showed a far better ammonia clearance capacity compared with our hybrid liver. Both modalities were cross-circulated with surgically induced liver failure dogs.

Regeneration of pancreas

We have previously established the bioartificial pancreas experimentally using the mouse insulinoma cell line, MIN-6, introduced by Prof Miyazaki from Osaka University. The MIN-6 was embedded into an immunoisolated, mesh-reinforced PVA hydrogel bag (MRPB). Each bag was implanted into the peritoneal cavity of DM rats, induced by strepotozotocin (55 mg/kg). The blood sugar of the DM rats was carefully controlled by this device minimally up to 3 months. Therefore, this kind of device would be applicable for clinical use if a large-scale isolation procedure for viable islets is completed.

It is particularly difficult to isolate the porcine islets compared with islets from other animals. In this study, we developed a stable procedure for adult porcine pancreatic islet isolation. After warm digestion in a water bath at 37°C for 45 min without mechanical shaking in the laboratory, the islets were purified with a COBE2991 cell processor on Euro-Collins Dextran T70 discontinuous density gradients (d = 1.108, d = 1.094, d = 1.081, Euro-Collins) by centrifugation at 2,000 rpm for 8 min. After purification, islets were collected and washed twice with cold Euro-Collins solution and 3 times with RPMI-1066. The number of harvested islets was about 85,900 islets/pancreas and about 1,694 islets/g. The islet equivalent was about 291,667 IEQ/pancreas and about 3,294 IEQ/g (pancreatic tissue). The purity was >90% and the success rate of adult porcine pancreatic islet isolation was raised from the original 10 to about 78% with our improved method. About 10,000 islets were used in the static incubation study without being hand-picked. Insulin secretion ranged from about 83 μU/dish/h in response to incubation with 3.3 mM of glucose. Insulin secretion was significantly increased to 104 μU/dish/h with stimulation at a higher concentration of glucose (16.7 mM). The concentration of insulin that was released in the culture medium was 83.19 mU/dish/24h (from about 10,000 islets without hand-picking) on the first day and 109.2 mU/dish/24h on the 2nd day in the same sample.

These islets will be mixed with 5% agarose and embedded into an immunoisolated bag and each MRPB will be implanted into the peritoneal cavity of DM rats that were induced using stropotozotocin (55 mg/kg).

Tissue Engineering for Therapeutic Use 1.
Y. Ikada and Y. Yamaoka, editors.

Tissue engineering for hard tissue

Project Leader

- Shoji Enomoto, Professor, Second Department of Oral and Maxillofacial Surgery, Faculty of Dentistry, Tokyo Medical and Dental University

Skeletal defects, whether from traumatic, degenerative, or tumorous causes, represent a major challenge to plastic and reconstructive surgery. Modern techniques for reconstructing these defects rely on the use of autologous bone and cartilage including nonvasculized or vasculized bone grafts, or implantation of synthetic prostheses. However, autogenous grafts are associated with problems due to the limited availability of donor tissue, donor site morbidity, complexity of the surgery, and inability to shape the graft to the exact configuration of the complex defects. Synthetic prostheses suffer from increased susceptibility to infection, incidence of extrusion, and uncertain long term interaction with the patients.

To resolve these problems, Vaccanti and Langer have demonstrated a new technology called tissue engineering that involves the morphogenesis of new tissue from constructs formed from isolated cells and biocompatible polymers. In addition, recent breakthroughs in molecular and cellular biology have made the regulation of gene expression with an array of growth and differentiation factors possible. In particular, the functions of many growth factors in growth and development are being made clear. However, few of these cell biology advances are being incorporated into the surgical armamentarium. In an attempt to advance our reconstructive capabilities, we attempted to unite the ideas and technology of surgery, engineering, and cell biology to create a new technology for reconstructing skeletal deformities.

The project team for tissue engineering for hard tissue consists of the following four subgroups; molecular biological research, orthopedic research, periodontal research, and maxillofacial reconstructive research. Each research group works in cooperation to explicate the basic mechanisms of hard tissue morphogenesis and regeneration, and to severally develop a unique and pertinent technique for regenerating hard tissue in each field.

Molecular biological research

Replacement of bone and cartilage defects in skeletal diseases and oral disorders with bioengineered tissues is one the final goals of the hard tissue project team. To support the efforts of these clinical research programs, this subgroup is involved in the elucidation of molecular and cellular mechanisms that are underlying hard tissue regeneration. We especially focus on the points described below:

1. The molecular mechanisms of the differentiation of osteoblasts, chondrocytes and osteoclasts.
2. Molecular events taking place in the repair process of fractures of bone and cartilage as well as in injuries to tendons and ligaments.
3. Cellular and molecular mechanisms operating in the bone cell responses to mechanical stimuli.

Cellular differentiation is based on activation of the expression of tissue-specific genes which are under the control of crucial tissue-specific transcription factors. Skeletal cells are derived from mesenchymal cells which give rise to bone and cartilage as well as fat, muscle, ligaments and tendons. Transcription factors are internal determinants of the process of cell differentiation while functions of bone and cartilage cells are also under the strict control of the extracellular signals including systemic and local regulators as well as extracellular matrices. Our final goal is to elucidate the transcription factors involved in tissue-specific gene expression and the factors affecting cell functions as communicating signals between cells and tissues.

Orthopedic research

The research of this team is focused on fracture repair and reconstruction of large bone defects, which require a strong structure. Basic studies for understanding the mechanism of bone healing and clinical experiments for the development of structurally strong implants are described below.

Bone-specific gene expression during bone healing

For osteogenesis to occur, mesenchymal cells must be committed to chondrogenic and osteogenic cell lineage. Cell-to-cell signaling peptides, called growth factors, which are thought to regulate chondrogenic and osteogenic commitment are abundant in bone matrix. We are studying the genes which encode information about growth factors and their receptors by in situ hybridization. Although the biochemical and regulatory aspects remain uncertain, large arrays of cartilage and bone-specific genes must be activated for bone to be formed. If their regulatory action can be described, growth factors might be used to accelerate fracture repair.

Bone filling materials for fractures and for bone defects

There are two new implants for stimulating bone regeneration in defects in long bones that are being studied. One is a titanium cage filled with PGS and bone morphogenetic proteins (BMP) and the other is a titanium mesh block soaked with high molecular weight hyaluronic acid and BMP. The implants are structurally strong and can each carry more than 2000 N of compressive force. The implants are placed into a defect of the canine femur. After sacrifice of the animals, the bones are biomechanically tested for strength and the morphology of the regenerated bone within the defect region is examined.

Periodontal research

The purpose of this project team is to establish a methodology for regeneration of periodontal tissues which were destroyed by periodontal diseases. Our research projects are mainly divided into the two groups described below.

In vivo studies

We are investigating alveolar bone, cementum and periodontal ligament regeneration with the use of growth factors, such as recombinant human bone morphogenetic protein-2 (rhBMP-2) in dogs, monkeys and rats. In concrete terms, a round through-and-through bone defect is created in the angular region of the rat mandible on both sides of the jaw with a steal round bur. rhBMP-2 with a polylactic acid-polyglycolic acid copolymer-gelatin sponge (PGS) is applied in the bone defect, then new bone formation is evaluated histometrically. We are also evaluating the effect of rhBMP in periodontal bone defects created in dogs or monkeys.

In vitro studies

We are also investigating cell differentiation mechanisms in periodontal regeneration induced by growth factors. We found gene expression of growth differentiation factor (GDF) -5, -6, and -7, in dental follicles from which periodontal tooth supporting tissues are developed. GDFs possibly have a significant role not only in the development but also in the regeneration of periodontal tissues such as alveolar bone proper, periodontal ligaments and cementum. On the other hand, it has been speculated that cells in periodontal ligaments (PDL) play an important role in the regeneration of the dental attachment apparatus because of their ability to form calcified tissues. We cloned two S100-calcium binding protein cDNAs from a bovine PDL library. We found that they are highly expressed in PDL of erupted teeth when compared to unerupted teeth, indicating that mechanical stress controls the expression of these calcium binding proteins.

Maxillofacial reconstruction research

The correction of cranio-maxillofacial contour irregularities and structural defects has the particularity compared to the reconstruction of long bones. The most critical demand is an aesthetic requirement. It demands that we not only regenerate bone tissue but also shape it to the exact configuration of the defects. The other difficulty is that reconstruction of the jaw bone is susceptible to infection because the bone is covered only with a thin mucous membrane. We also have to take into account the rehabilitation of specific oral functions such as speech and mastication. This team aims to establish a methodology for reconstruction of maxillofacial deformities to fulfill the above mentioned requirements.

It is necessary for tissue regeneration to control three important elements; cells, matrices, and growth and differentiation factors. In addition to these three elements, molding of the shape is important for controlling the configuration of regenerated bone.

We are studying each element separately and testing new methods by integrating these four elements into animal experiments. Specifically, we focus on the points described below.

Cells

The most noticeable cell for bone regeneration is the bone marrow cell because it is easy to harvest and has the potential to differentiate into an osteoblast. To evaluate the effect of growth factors on bone marrow cells, cells are cultured with various growth factors such as TGF-b, FGF and BMP and several markers are examined for cell growth and osteoblastic differentiation.

Matrix

As a scaffold to sustain cells, we use both synthetic polymers and ceramics. Materials with various compositions and configurations are prepared. Cells are seeded into these materials with or without growth factors, and the osteoinductive potential is examined both in vivo and in vitro. We also evaluate the effect of natural matrices such as collagen and proteoglycan.

Growth and differentiation factors

Among the various growth factors, we are interested in the TGF-b superfamily, especially in BMP in terms of bone induction. Besides the experiment for testing the effects of BMP in vitro and in vivo, we are studying the receptors and the signaling molecule of BMP to understand the mechanism of the action of BMP. We are analyzing activin-like kines (ALK) as specific receptors for each BMP, and MAD-1 as a messenger of BMP signals.

Animal experiments

The implants, which consisted of synthetic matrices, growth factors, and cells, are implanted into the subcutaneous tissue or calvarial bone defects of rats. These implants are also used to reconstruct segmental bone defects of a monkey mandible with synthetic polymer or a titanium mesh tray as a mold. The bone inducing potential of these implants is evaluated histologically.

Contents

Tissue Engineering for Therapeutic Use 1.
Y. Ikada and Y. Yamaoka, editors.

Tissue engineering research trends at Kyoto University

Yoshito Ikada
Research Center for Biomedical Engineering, Kyoto University, Kyoto, Japan

Keywords: growth factors, Kyoto University, organ substitution, tissue regeneration, tissue engineering.

Introduction

Tissue engineering is a new technology that makes use of living cells for regeneration of natural tissues and for organ substitution by bioartificial organs. Tissue regeneration is performed either in vivo or in vitro. Cells regenerating new tissue are seeded on a scaffold in the case of in vitro tissue regeneration, but cell seeding on a scaffold is not always necessary when a new tissue is regenerated in vivo, because cells will infiltrate the implanted scaffold from the surrounding living tissue. Unless cells are seeded, cell growth factors are usually incorporated into the scaffold to promote tissue regeneration. Scaffolds used for tissue regeneration are normally bioresorbable. Biodegradable sheets and tubes are also employed to make space for tissue regeneration and simultaneously to prevent the ingrowth of fibrous collagenous tissue into the defective site.

If a tissue that is to be replaced using tissue engineering technology should execute highly metabolic functions, as do the liver and pancreas, living cells possessing the metabolic functions are hybridized with nonbiodegradable polymers that act as immunoisolative membranes. The hybrid-type artificial organs are of temporary or permanent use for organ substitution.

Tissue regeneration

A variety of tissues are the targets of current tissue engineering research at Kyoto University. Major tissues studied at Kyoto University are briefly described below.

Scaffold alone

Skin (dermis) [1,2]
Though various kinds of skin equivalents have been developed to cover skin defects resulting from burns or trauma, most of them have served only as a tem-

Address for correspondence: Yoshito Ikada, Research Center for Biomedical Engineering, Kyoto University, 53 Kawahara-cho, Shogoin, Sakyo-ku, Kyoto 606-01, Japan. Tel.: +81-75-751-4115. Fax: +81-75-761-4144.

porary skin substitute. By modifying the technique proposed by Yannas, Burke et al. we developed a new "bilayer artificial skin" composed of an upper silicone sheet and a lower sheet of collagen sponge containing chondroitin-6-sulphate, a type of glycosaminoglycan (GAG). When the "bilayer artificial skin" is applied to full thickness skin defects, cellular tufts of fibroblasts and capillaries infiltrate the collagen sponge layer; the lower sponge layer then spontaneously converts to a synthesized connective tissue matrix similar to dermis in 2—3 weeks. Secondary thin split thickness skin grafts placed on the dermis-like tissue take easily and postoperative graft contraction is minimal.

After our clinical report on the original version was published, an experimental study revealed that there was no significant improvement achieved by adding GAG to the collagen sponge sheet. Therefore, a revised version of the "bilayer artificial skin" without GAG was produced for clinical use to reduce the primary cost. Preliminary use of this prototype without GAG produced results equal to that of the original version with chondroitin-6-sulphate. Since our early results from the experimental and clinical use of the original version of the "bilayer artificial skin" were reported, several improvements have been made in stages to eliminate some drawbacks related to disinfection and preservation and to reduce the primary cost of manufacture. The latest version of the material was successfully used in 27 sites on 23 patients.

Dura mater [3]

Various natural and synthetic substitutes, including autografts, allografts, xenografts, and nonabsorbable and absorbable synthetic polymer sheets, have been used to repair a defect of dura mater in both experimental models and clinical practice. Although an autograft, such as fascia and pericranium, does not induce immunological or severe inflammatory reactions, a sufficiently sized autograft is hard to harvest from a patient. Therefore, allografts such as LYODURA® have been used most widely in clinical practice, but they are not completely free from a potential risk of latent virus infection. Very recently the LYODURA® supplier stopped to sell this product in the world because of shortage of the allogeneic donors. As a result, nonabsorbable synthetic materials such as Silastic® and GoreTex® are obliged to be used as a dural substitute, but Silastic® is reported to cause a chronic inflammatory reaction inducing long-term delayed intracapsular hemorrhage from the unusual vascularity in the connective tissue formed. Also, bioabsorbable materials such as collagen film, elastin-fibrin derivative, Vicryl® mesh, and a composite sheet of Vicryl® mesh and collagen, have been examined as dural substitute. Although these bioabsorbable materials do not induce a severe chronic inflammatory reaction, their degradation rate is not well controlled, so that they tend to disappear before regeneration of a strong connective tissue. An ideal dural substitute must be strong and flexible, and can be tightly sutured to the edge of defects of dura mater. In addition, it must be absorbed under gradual replacement by a connective tissue forming a dura-like tissue.

To develop such a biomaterial, we fabricated a new bioabsorbable composite sheet. This bioabsorbable composite, composed of two L-lactide-ε-caprolactone (50/50) copolymer films and a poly(glycolide) nonwoven fabric, had good mechanical properties and was completely absorbed 24 weeks after implantation in the back of rats. Histological evaluation of the composite sheet was undertaken by implanting it in 31 rabbit dural defects from 2 weeks to 26 months. Neither infection, cerebrospinal fluid leakage, evidence of convulsive disorders, significant adhesion to the underlying cortex, nor calcification were noticed in any of the cases. In addition, the regenerated dura-like tissue had a high pressure-resistant strength 2 weeks after implantation. It was concluded that this new bioabsorbable composite sheet could be successfully used as a dural substitute.

Periodontal tissue [4,5]

We investigated regeneration of osseous defects in dogs following placement of a lactic acid-glycolic acid copolymer (PLGA) membrane around the roots of treated teeth and compared the healing with that of the control teeth without the membrane. Mandibular premolar teeth of five dogs were used for experimentation. Specimen blocks were removed 8 weeks postoperatively for histological evaluation of contralateral experimental and control teeth. It was found that the mean apical migration of junctional epithelium along the experimental surface amounted to 0.25 mm as compared to 0.39 mm for the control surface and the mean formation of new cementum along the experimental surface amounted to 1.68 mm as compared to 0.41 mm for the control surface. Mean formation of new bone along the experimental surface was found to amount to 0.84 mm as compared to 0.46 mm for the control surface. Neither ankylosis nor severe resorption was observed on the experimental surface. It was concluded that the PLGA membrane placement was effective in inhibiting epithelial migration and encouraging formation of new connective tissue attachment to the root surface.

The clinical efficacy of this membrane for guided tissue regeneration (GTR) was assessed. A total of 60 patients with mainly advanced adult periodontitis, including lesions with angular bony defects, two-wall, three-wall and compound infrabony defects and class I, class II, and class III furcation defects, were studied. The group included 22 males and 38 females: ages ranged from 24 to 68 years with a mean age of 48 years. Following completion of the hygienic phase of treatment, clinical measurements including probing depth and clinical attachment level were recorded. Bone defects were surgically exposed. Defects were treated with open flap debridement plus absorbable periodontal membrane. Postsurgery and clinical data were recorded for 6 months. Postsurgical healing was uneventful. Inflammation of the soft tissue covering the device or other adverse events related to this absorbable membrane have not been noticed in any patients. The mean probing depth was reduced from 5.7 to 2.9 mm, with a mean probing depth reduction of 2.8 mm ($p<0.001$). Relative attachment level was improved from an average of 9.0 to 6.8 mm, with a mean gain of probing attachment of 2.2 mm ($p<0.001$). There was minimal gingival recession, averaging 0.6 mm. These results suggest that the absorbable membrane used in the GTR procedure is valuable in the treatment of adult periodontitis.

Ligament [6,7]
Autogenous and synthetic materials have been used in reconstruction surgery of the ruptured anterior cruciate ligament (ACL). Autogenous tissues such as semitendinosus tendon, patellar tendon, fascia lata etc., have been used as graft materials in ACL reconstruction. However, it was found that these tissue transplants underwent an initial loss in mechanical properties followed by a period of revascularization and remodeling. During this period, their mechanical resistance reduced. If excessive stress is applied before the tissue has matured, the autogenous tissue may rupture or elongate. Therefore, it is necessary to protect the grafts by some immobilization, but it is well known that immobilization also leads to loss of ligament strength. In order to overcome this paradox, the ligament augmentation device (LAD) was used in ACL reconstruction. The use of an LAD made of polypropylene (PP) braid (3M, St. Paul, Minnesota, USA), has been a mechanism to protect healing autogenous patellar-ligament reconstruction in early postoperative period and was proposed by Kennedy et al. Kennedy-LAD™ was not designed to be a permanent prosthetic replacement for ACL, but designed so that load-sharing between the LAD and the biological graft would protect the graft during the period of degeneration and weakening. The LAD would eventually deform and the load transfers to the biological graft even if the LAD is left permanently within the joint. However, because the material is PP which is nonresorbable, there is some risk of inducing long-term stress shielding of the augmented graft. In order to avoid the potential long-term stress shielding and to ensure progressive full load transfer to graft tissue, bioresorbable poly(L-lactide) (PLLA)-LAD has been developed. PLLA is a synthetic, biodegradable linear polymer and is already used clinically in different applications such as maxillofacial and mandibular surgery, dentistry, and orthopaedics. In vivo, a great advantage of the PLLA is the degradation by simple hydrolysis of the ester bond of the polymer chain and the degradation products are finally metabolized to carbon dioxide and water which are excreted by the lungs and kidneys. The degradation of PLLA is completed in 10 months to 4 years depending on the molecular weight, shape, and size of the implants. We fabricated PLLA-LAD with the same design as the conventional Kennedy-LAD™ and determined the efficacy of bioresorbable LAD by comparing the physical, biomechanical, and histological properties of PLLA-LAD with those of conventional nonresorbable PP-LAD. The initial physical properties of PLLA-LAD were almost the same as those of PP-LAD. Biomechanics and tissue response to the PLLA-LAD were examined by using bone-patella tendon-bone (BTB) for ACL reconstruction in goat model up to 2 years. PLLA-LAD induced no deleterious clinical and histological effects in this period.

Simultaneous reconstruction of anterior and posterior cruciate ligaments (ACL and PCL) of knee joints was clinically attempted for three patients using the

PLLA-LAD described above. A 9 mm-wide LAD was covered with autogenous iliotibial ligament for ACL regeneration, while autogenous patellar tendon with a connected bone was rolled around a 5 mm-wide LAD for PCL regeneration. The LAD for ACL was passed through a bone tunnel 7 mm in diameter and fixed using the double stapling technique. On the other hand, both ends of bone pieces on the patellar tendon were fixed with interference screws under full extension of the patellar tendon and PLLA-LAD, after passing through a bone tunnel 12 mm in diameter. The ends of LAD were then fixed by double stapling technique. Observation for 1 year and 2 months after the operation on average showed that the anterior/posterior laxity was well controlled without any articular inflammation or blood abnormality.

Esophagus [8]
Various types of artificial esophagi have been designed aiming to reduce surgical complications after esophageal replacement. However, complications, such as leakage, infection, stenosis, and dislocation, have remained inevitable, and therefore, these devices have proven to be far from useful in clinical practice. These earlier designs were made of natural substances, synthetics, or composites. However, even if an artificial esophagus has a high affinity for recipient tissue, it remains a foreign body to the host and will not be able to overcome these complications. We developed an artificial esophagus, with which problems of infection, anastomotic leakage, stricture formation, and dislocation of prosthesis were resolved. A new type of artificial esophagus that we have developed consists of an inner silicone tube and an outer nonantigenic collagen tube. The novel feature of this artificial esophagus is that the main part of the prosthesis is replaced by host tissue. We have already found that no stenosis of artificial esophagus developed when the replacement part was stented for more than 4 weeks. It was considered that this stenosis was caused mainly by poor regeneration of submucosal tissue, rather than by the grade of reepithelialization. Recently, we found that in cases in which the stent dropped out within 3 weeks, fibrous tissue was noted beneath the neoesophageal epithelium. In such a case, neither muscle layers nor submucosal glands regenerated beneath the neoesophageal epithelium after replacement. However, in cases in which the stent dropped out more than 4 weeks after surgery, the neoesophagus was covered with a polylayer of squamous epithelium and had normal esophageal glands and a muscle layer. Therefore, we concluded that muscle tissue and esophageal glands were able to regrow in the neoesophagus when the portion replaced by the artificial esophagus was stented for at least 4 weeks.

Peripheral nerve [9,10]
To evaluate the regeneration of resected nerves, a dehydrothermally treated gelatin tube was used to repair a 10-mm defect in the sciatic nerve of rats. As a control, a gelatin tube treated with glutaraldehyde was used. Histological examination using osmium, Bodian and anti-S-100 protein staining at 2 and 4 months

after operation revealed that the regenerated axons reached the distal end through the gelatin guide-tube. Motoneurons of the ventral horn as well as afferent nerve terminals in the dorsal horn of the spine were stained both retrogradely and anterogradely with WGA-HRP, indicating nerve regeneration through the bridged area. With regard to nerve functions, the electromyogram and cerebral somatosensory potentials at 2 months showed functional recovery. In the controls, no regeneration was observed.

We have also examined the nerve regeneration that occurs over a 25-mm gap using a novel biodegradable nerve guide tube. The tube was a composite of polyglycolic acid (PGA) mesh coated with collagen which was filed with neurotrophic factors. The left sciatic nerve of 10 adult cats was dissected. The stumps were connected by the tube and fixed gap. Histological examinations carried out 4–16 months after implantation of the tube revealed regeneration of well vascularized nerve tissue. Regeneration of both myelinated, unmyelinated axons and Schwann cells was confirmed by electron microscopy 5 months after surgery. Following an injection of horseradish peroxidase (HRP) into a site peripheral to the regenerated segment of sciatic nerves, motoneurons in the ventral horn of the spinal cord, afferent terminals in the medial portion of the dorsal column of the medulla oblongata, and sensory afferent nerve terminals in the dorsal horn of the spinal cord were labelled. Electrophysiological examinations revealed restoration of evoked electromyograms and sensory evoked potentials (SEPs) recorded from the cerebral cortex as well as the spinal cord. It was also found that some regenerated motor axons exhibited branching in regenerated segments. In two cases, a single motoneuronal axon from the regenerated side projected to both flexors and extensors simultaneously. These results indicate that the PGA-collagen composite tube is a promising tool for use as a nerve guide tube in peripheral nerve regeneration.

Scaffold and growth factor

Capillaries and arterioles [11]

To regenerate capillaries and arterioles, i.e., to promote neovascularization, we designed a drug formulation of basic fibroblast growth factor (bFGF) composed of acidic gelatin with the isoelectric point of 4.9 on the basis of its property of forming polyion complex with basic bFGF molecules. This proposed hydrogel formulation mimics the manner in which bFGF is stored in the extracellular matrix. We prepared a biodegradable hydrogel by cross-linking acidic gelatin with water-soluble carbodiimide. In this hydrogel system, the degradation period of hydrogels varied depending on their cross-linking extent, which could be regulated by altering the concentration of gelatin and the carbodiimide used for hydrogel preparation. Release experiments demonstrated that bFGF was not released from gelatin hydrogel under in vitro nondegradation conditions. The hydrogels degraded with time following subcutaneous implantation on the mouse back. The period of in vivo degradation varied depending on the water content of hydrogels. Gelatin hydrogels containing bFGF strongly induced neovascularization and tissue granulation around the implantation site, whereas subcutaneous injections of bFGF in solution form did not increase the amount of tissue hemoglobin or wet weight of tissue. The bFGF-impregnated gelatin hydrogel of lower water content retained the vascularization effect for a longer period. It was concluded that bFGF was released from gelatin hydrogel as a result of their in vivo degradation, leading to more efficient induction of neovascularization and tissue granulation than free bFGF without impregnation in hydrogel.

Skull bone [12]

Various biomaterials have been used to reconstruct skull bone defects in neurosurgery, including autografts with or without vessels, titanium, methyl methacrylate resins and bioactive ceramics. However, many problems remain which need to be addressed before the successful clinical application of these materials to skull defects can be achieved. For example, autografts are often associated with donor-site pain and high morbidity and infection rates, in addition to limited supplies of graft materials. Other materials have been used only as a cap for defects because they are not capable of inducing bone formation. If a material that has osteoinductive activity is found, it would open a new route for bone reconstructive surgery by facilitating reconstruction of compromised skull bone and sinus defects in skull base surgery.

It has been reported that bFGF can stimulate bone formation, but that when injected in free form, it does not stay at the injection site long enough for its effective in vivo biological activity to produce the expected results. In an attempt to overcome this problem, we have developed a biodegradable hydrogel composed of acidic gelatin to enable bFGF to be released at the site of action for an extended time period. The bFGF tends to be ionically bound to the acidic gelatin and is released from the matrix when the gelatin degrades. We have previously demonstrated that incorporation of bFGF in hydrogel enhances the in vivo angiogenetic effect of bFGF, in marked contrast with free bFGF.

We evaluated the biodegradable gelatin hydrogels incorporating bFGF for efficacy in bone regeneration using a rabbit model. Hydrogels with water contents of 85 and 98% were prepared using chemical cross-linking of gelatin with an isoelectric point of 4.9 in aqueous solution and, after freezedrying, were impregnated with an aqueous solution of bFGF to obtain bFGF-incorporated gelatin hydrogels. When they were implanted into bone defects measuring 6 mm in diameter in rabbit skulls (six animals/group), complete closure of the defect was observed at 12 weeks after implantation, regardless of the water content of the hydrogels. In contrast, bFGF did not enhance bone regeneration when applied to the skull defect in solution with phosphate-buffered saline (PBS). Furthermore, gelatin hydrogels lacking bFGF were not effective in inducing bone formation, with fibrous tissue growing into the defect instead, similar to the skull defect seen in control rabbits treated with PBS. This indicates that presence of hydrogels did not interfere with bone regeneration at the skull defect, probably because of their disappearance during biodegradation. It was concluded that gelatin hydrogel is a promising matrix for effective induction of biological activity of bFGF for bone regeneration in skull and sinus defects.

Bone (ectopic) [13]
Bone regeneration has attracted much attention in the field of tissue engineering because of its high level of clinical demand. To regenerate the patient's own bone tissue in osseous defects, a number of methods have been proposed, including space making with a membrane, especially in maxillofacial surgery for guided bone regeneration, and scaffolding for bone ingrowth in orthopedic, plastic and skull-base surgeries. Scaffolds employed in bone regeneration basically function as the substrate for bone cell attachment and, therefore, are made of porous biodegradable materials such as glycolide-lactide copolymer nonwoven fabrics, collagen-sponges and porous calcium phosphate blocks. On the other hand, biodegradable matrices incorporating cell growth factors such as bone morphogenetic protein (BMP), transforming growth factor-β, and FGF have been placed in bone defects or implanted at an ectopic site. In this case the matrices serve as the carrier or depot of the growth factors and occasionally function also as the substrate for bone cell attachment. Fibrin glue, glycolide-lactide copolymers and calcium phosphates have been used as matrix of bone-related growth factors.

We have studied the effect of the shape of carriers on bone regeneration. Two biodegradable polymeric materials, a polyglycolide (PGA) nonwoven fabric and a gelatin hydrogel, were used as carriers of recombinant human bone morphogenetic protein-2 (rhBMP-2). The PGA nonwoven fabric was made from PGA fibers of 20 μm diameter without using any binders, while the gelatin hydrogel was prepared by cross-linking gelatin in aqueous solution with glutaraldehyde to a water content of 95% when swollen with water. Following impregnation of rhBMP-2, the carriers of disk-type were implanted into Wistar rat thigh muscle. Induction of ectopic bone formation from rhBMP-2-impregnated carriers was evaluated by Softex and histologic observation, after staining the explanted tissue with alizarin red S stain to identify calcium deposition. Both of the biodegradable polymeric carriers containing 10 μg of rhBMP-2-induced ectopic bone formation after 2 weeks of implantation but not after the 1st week. A remarkable finding was a difference in the macroscopic morphology between the ectopic bones induced by PGA nonwoven fabric and the gelatin hydrogel. The PGA nonwoven fabric containing rhBMP-2 induced ectopic bone formation inside the carrier, whereas the gelatin hydrogel formed bone only at the periphery of the carrier.

Scaffold + cells

Trachea [14]
Studies of prosthesis for repairing tracheal defects have a long history, beginning with Daniels' experiment in 1948. Since then, a great variety of tubular prosthesis

have been used, but no ideal graft technique has been established. We have been studying a porous tracheal prosthesis that is highly biocompatible, and have developed an innovative tracheal prosthesis made from collagen grafted mesh. Using this prosthesis, we performed intrathoracic tracheal replacement in dogs, with a view to possible clinical application. The prosthesis consists of Marlex mesh (polypropylene) reinforced with a continuous polypropylene spiral, and is grafted and coated with pig collagen (types I and III). Complete surgical resection of the mediastinal trachea was performed in seven adult mongrel dogs. In one dog, a four-tracheal ring segment (2 cm) was resected and replaced with a 3-cm prosthesis, and in six dogs, a seven- to eight-tracheal ring segment (4 cm) was resected and replaced with a 5-cm prosthesis. In the latter six dogs, a silicone tube was temporarily inserted into the replacement, and removed by bronchoscopy 1 month after surgery. In one dog that received a 4-cm replacement, we added omentopexy around the reconstructed trachea. The prostheses in all dogs were promptly infiltrated by surrounding tissue and incorporated by the host trachea. No dehiscence or air leakage was observed after surgery. Mild luminal stenosis was evident in one dog, and partial exposure of the mesh (unceration) was observed in five dogs within an observation period of 3—26 months. However, in the dog that received omentopexy after tracheal reconstruction, no stenosis or ulceration was observed, and the luminal surface seemed lustrous even after 6 weeks. Formation of respiratory epithelium, which lined the prosthetic lumen, was seen to various degrees: in the two dogs killed 12 and 26 months after surgery, confluent epithelization was confirmed histologically from the upper to the lower anastomosis of the prosthesis. The tracheal prosthesis is useful for the repair of intrathoracic tracheal defect and shows promise for clinical application with further investigation.

Scaffold + cells + growth factor

Cartilage (fibrous) [15]
Cartilage replacement is clinically needed, especially in maxillofacial, orthopaedic and plastic surgery. It is mostly silicone prostheses and autologous rib bones that have been used for this purpose, but several problems are involved in these cartilage replacements, such as infection at the interface between the implanted material and the tissue, and deterioration of the donor site by filing with fibrous cartilage tissue. Therefore, a preferable replacement would be to use the natural cartilage tissue. The first attempt of using the cultured chondrocytes for articular cartilage repair was reported by Green and it is supposed that a template is necessary for cartilage reconstruction by chondrocytes in vivo. There are some reports describing the seeding of collagen and other porous matrices with chondrocytes for this purpose. Italy et al. reported the repair of bone tissue by chondrocytes incorporated in collagen gel as a template. Langer, Vacanti, and their co-workers have developed a tissue engineering technique with the use of chondrocytes of the target region. The cells are isolated from patients, proliferated by

the in vitro culture to a higher density and then implanted after preparing a chondrocyte-polymer composite as the core for the cartilage reconstruction. It is reported that chondrocytes can be isolated from tissue and grown in culture in such a way as to maintain their phenotype.

We have prepared a chondrocyte-collagen composite in an attempt to regenerate cartilage by its subcutaneous implantation in nude mice. When the composite was impregnated with bFGF prior to implantation, regeneration of the cartilage tissue was remarkably accelerated. Histological staining of the implanted composite with Safranin O-fast green revealed that the cells incorporated in the composite exhibited their phenotype and formed a new matured cartilage. A thin layer of fibrous capsule was observed surrounding the implanted composite and the inflammatory response of the host to the implant was mild. Specific proteoglycans had accumulated in the composite even 1 week after implantation. At 2 weeks postimplantation the chondrocytes regenerated the cartilage tissue, although still immature, but at 4 weeks almost all of the chondrocytes had transferred to the mature stage. Conversely, such mature cartilage tissue was not noticed for up to 4 weeks after implantation if the collagen scaffold was not impregnated with bFGF. Moreover, the mature area was limited to only a small fraction of the implanted composite, unless bFGF was incorporated in it.

Jaw bone [16]

The loss of the mandibular arch after tumor excisions or traumatic injuries not only leaves the patient with a conspicuous deformity of the face, but also with a mandible that does not function for chewing, swallowing, or speaking. Even if the patient becomes free of disease, these deformities may inflict physical and mental pain and make social rehabilitation difficult. So far, a variety of methods and materials to reconstruct defective mandibles are available: transplanting of autogenous or allogeneic bone grafts, implantation of metal plates and artificial bones from ceramics. Although these methods have advantages and disadvantages, reconstruction of the defects of the mandible remains a difficult challenge for oral and maxillofacial surgeons. The goal should be to reconstruct the jaw so that the patient can perform normal functions. To achieve this, it is essential that a method of promoting regeneration of physiological bone is developed.

Recently, attention has been paid to jaw bone reconstruction involving transplantation of autogenec particulate cancellous bone and marrow (PCBM). PCBM contains abundant amounts of osteoblasts, their progenitor cells, and bone matrix. Therefore, it has a high bone formation capability. Also, it allows freedom in shaping an implant into a required shape and can be secured repeatedly from the same donor site. However, in using PCBM to reconstruct the jaw, a tray must be used to maintain the grafts in an appropriate form. In addition, the prosthesis must be able to assist the newly formed bone in acquiring the capability to withstand external forces and to promote diffusion of nutrients to the grafted tissue from the recipient bed. Frozen, heat-treated autogenous cortical bone trays, metallic trays, and Dacron-polyurethane mesh trays have been used for this purpose. The frozen, heat-treated autogenous cortical bone is complicated to use and can rarely be applied in the case of severe bone destruction. A metallic tray is easily exposed through the skin and vulnerable to infection, and its stiffness causes atrophy in the transplanted bone. On the other hand, the Dacron-polyurethane mesh tray has affinity for the living body tissue, malleability and appropriate elasticity to the bone. However, its nonbiodegradability, similar to metallic trays, causes a foreign body reaction as long as the material is present within the body. To solve these problems by reconstruction of mandibular continuity defects using fresh autogenous PCBM, we have designed a biodegradable tray for this application using a poly(L-lactide) (PLLA) mesh. This tray will make physiological reconstruction of the jaw possible. In the future this method is also expected to assist in the development of simultaneous functional reconstruction of the jaw bone and oral rehabilitation when dental implants are included in the tray.

Organ substitution

Temporary use

Liver assist device [17]
We have been studying a bioartificial liver assist device comprised of hepatocytes within a hollow fiber module (system I) and an extracorporeal circulation system including a swine whole liver (system II) for potential therapy of reversible hepatic injury or for bridge use in liver transplantation.

System I. Hepatocytes were collected from a young pig weighing 15 kg by the in situ collagenase perfusion method. About 10^{10} hepatocytes ($\sim$100 g) suspended in culture medium without any polymeric additives, such as collagen, were inoculated into the lumen of hollow fibers with a nominal 30 kDa molecular cutoff. A mixture of human blood (type A) and DMEM culture medium (1:4) was circulated in the outer space at 200 ml/min. Ammonia, galactose, and lidocaine solutions were applied to the circulating medium and these intrinsic clearances were determined from the decrease of the concentrations to evaluate the functions of the device.

System II. A whole liver ($\sim$420 g) isolated from a young pig was employed as a bioreactor. For an in vitro study human whole blood (type A) was circulated through the portal vein at 400 ml/min and through the liver artery at 100 ml/min. The blood was supplemented with 25 mg/h of soluble form PE1. The intrinsic clearance of ammonia, galactose and lidocaine was determined. Their intrinsic clearances are listed in Table 1.

The metabolic capacity of the hollow fiber module is 1/40 to 1/20 of that of the whole liver. Because hepatocytes are isolated from the host by a semipermeable membrane, the hepatocytes protected from immune rejection may maintain the metabolic functions for a long period of time. The risk of disease transmission from animals to the host will also be greatly reduced. Conversely, system II can afford a powerful liver assist, but its operation period is limited to about 15 h.

Permanent use

Bioartificial pancreas [18,19]
Although there has been increasing evidence that pancreatic islet transplantation may offer an ideal endocrine replacement therapy for patients with type I diabetes mellitus (DM), one of the major problems may be recurrence of the original disease in the unprotected islet cells. Immunosuppressive therapy is, at present, indispensable for the clinical application of islet transplantation in spite of its toxicity. A possible solution to these problems is encapsulation of donor islets in a biocompatible polymer capsule membrane that constitutes a physical barrier to the host's immune system. The need for any kind of immunosuppression might be eliminated even in the future clinical application of islet transplantation with the development of a unique immunoisolation system for islets. We have been studying a bioartificial pancreas with entrapment of pancreatic islets in poly(vinyl alcohol) membrane in rats with experimentally induced diabetes. Glucose, insulin and nutrients pass through the membrane easily, whereas the passage of immunoglobulin G is completely prevented, indicating that this membrane can be effective in protecting the bioartificial pancreas from immunorejection. Approximately 2,000 islets collected from three Sprague-Dawley rats were enclosed in a mesh-reinforced poly(vinyl alcohol) tube and transplanted into the peritoneal cavity of six Wistar rats with streptozotocin-induced diabetes. Their nonfasting serum glucose levels were significantly decreased for at least 12 days. Six diabetic rats receiving intraperitoneal transplantation of free islets without the tube showed a slight but significant decrease in nonfasting serum glucose levels for only 3 days. One diabetic rat with transplantation of the bioartificial pancreas had a significant and sustained decrease in nonfasting glucose levels from pretransplanted levels of 440—500 mg/dl to a mean value of 162 ± 13 mg/dl over 3 months without immunosuppression. The bioartificial pancreas was then removed and glucose levels gradually increased to over 500 mg/dl. This result suggests that a bioartificial pancreas with entrapment of islets in a poly(vinyl alcohol) membrane is a promising therapeutic approach to diabetes mellitus.

We are also studying an improved version of a three-layer agarose microcapsule in islet xenotransplantation. The microcapsule is composed of a mixture of 5% agarose and 5% polystyrene sulfonic acid. The other two outer layers are polybrene and carboxymethyl cellulose. The agarose/polystyrene sulfonic acid membrane is used for immunoisolation, suppression of complement activity and reinforcement of the microcapsule. The polybrene layer suppresses the polystyrene sulfonic acid leakage by forming a polyionic complex at the surface of the agarose/polystyrene sulfonic acid membrane. The outermost layer, a carboxymethyl cellulose coating, improves the biocompatibility of the microcapsule. In vitro static incubation studies showed that the insulin secretion from the rat islets in microcapsules in response to 16.7 mM glucose stimulation was 4 times higher than that on 3.3 mM glucose stimulation (n = 8). In an in vivo study, 500 rat islets in microcapsules were xenogeneically implanted in the abdominal cavity of mice with streptozotocin-induced diabetes. The graft survival times ranged from 2 to 5 months, the average being 75 days (n = 5). Our results demonstrate that the improved version of the three-layer agarose microcapsule can effectively prolong the xenograft survival time without employing immunosuppressants, suggesting that this microcapsule could provide a promising biohybrid artificial pancreas for future clinical applications.

References

1. Suzuki S, Matsuda K, Maruguci T, Nishimura Y, Ikada Y. Further applications of "bilayer artificial skin". Br J Plas Surg 1995;48:222–229.
2. Suzuki S, Matsuda K, Nishimura Y, Maruguchi Y, Maruguchi T, Ikada Y, Morita S, Morota K. Review of acellular and cellular artificial skins. Tis Eng 1996;2:267–275.
3. Yamada K, Miyamamoto S, Nagata I, Kikuchi H, Ikada Y, Iwata H, Yamamoto K. Development of a dural substitute from synthetic bioabsorbable polymers. J Neurosurg 1997;86:1012–1017.
4. Yamada S, Matsumoto Y, Takahashi Y, Yamanouchi K, Aoki H, Sato T, Ishikawa T, Hyon SH, Ikada Y. Histopathologic study of poly(lactic acid-co-glycolic acid) membranes to guided tissue regeneration in dogs. J Jpn Soc Periodontol 1991;33:396–405 (In Japanese).
5. Yamanouchi K, Nakagawa T, Seida K, Saito A, Yamada S, Hiwatashi K, Setoguchi T, Chuman M, Sueda T. Clinical study on the effect of absorbable membrane applied to guided tissue regeneration technique. J Jpn Soc Periodontol 1994;36:884–894 (In Japanese).
6. Kobayashi H, Kawamoto Y, Hara S, Tomizawa N, Gibbons DF, VanKampen CL, Mendenhall HV. Study on bioresorbable poly-L-lactide(PLLA) ligament augmentation device (LAD). Proceedings of the 4th Japan International SAMPE Symposium, September 1995. 1995:25–28.
7. Ishimura M, Ohgushi H, Inoue K, Fukuda T, Tamai S, Ikada Y, Tomita N, Fujisawa Y. Clinical experiences of simultaneous reconstruction of anterior and posterior cruciate ligaments of knee joint using an artificial ligament made of poly(L-lactide), Hiza (Knee) 1995;21:86–89 (In Japanese).
8. Takimoto Y, Teramachi M, Okumura N, Nakamura T, Shimizu Y. Relationship between stenting time and regeneration of neoesophageal submucosal tissue. ASAIO J 1994;40;M793–M797.
9. Lee G, Nakamura T, Shimizu Y, Tomihata K, Ikada Y, Endo K. Experimental study of a nerve guide-tube made from dehydrothermally treated gelatin – application to repair of gap in rat sciatic nerve. Jpn J Artif Organs 1993;22:364–369 (In Japanese).
10. Kiyotani T, Teramachi M, Takimoto Y, Nakamura T, Shimizu Y, Endo K. Nerve regeneration across a 25-mm gap bridged by a polyglycolic acid-collagen tube: a histological and electrophysiological evaluation of regenerated nerves. Brain Res 1996;740:66–74.
11. Tabata Y, Hijikata S, Ikada Y. Enhanced vascularization and tissue granulation by basic fibroblast growth factor impregnated in gelatin hydrogels. J Contr Rel 1994;31:189–199.
12. Yamada K, Tabata Y, Yamamoto K, Miyamoto S, Nagata I, Kikuchi H, Ikada Y. Potential efficacy of basic fibroblast growth factor incorporated in biodegradable hydrogels for skull bone regeneration. J Neurosurg 1997;86;871–875.
13. Yamamoto M, Kato K, Ikada Y. Effect of the structure of bone morphogenetic protein carriers on ectopic bone regeneration. Tis Eng 1996;2:315–326.
14. Okumura N, Teramachi M, Takimoto Y, Nakamura T, Ikada Y, Shimizu Y. Experimental reconstruction of the intrathoracic trachea using a new prosthesis made from collagen grafted mesh. ASAIO J 1994;40:M834–M839.

15. Fujisato T, Sajiki T, Liu Q, Ikada Y. Effect of basic fibroblast growth factor on cartilage regeneration in chondrocyte-seeded collagen sponge scaffold. Biomaterials 1996;17:155–162.
16. Kinoshita Y, Kobayashi M, Fukuoka S, Yokoya S, Ikada Y. Functional reconstruction of the jaw bones using poly(L-lactide) mesh and autogenic particulate cancellous bone and marrow. Tis Eng 1996;2:327–341.
17. Iwata H, Sajiki T, Park YG, Zhu B, Ito E, Ikada Y, Sato S, Yagi T, Kanazawa A, Uesugi T, Yoneyama T, Ikai I, Yamaoka Y. Preparation of a bioartificial liver and quantitative evaluation of its metabolic functions. ASAIO J 1997;(In press).
18. Inoue K, Fujisato T, Gu YJ, Burczak K, Sumi S, Kogire M, Tobe T, Uchida K, Nakai I, Maetani S, Ikada Y. Experimental hybrid islet transplantation: application of polyvinyl alcohol membrane for entrapment of islets. Pancreas 1992;7:562–568.
19. Tun T, Inoue K, Hayashi H, Aung T, Gu YJ, Doi R, Kaji H, Echigo Y, Wang WJ, Setoyama H, Imamura M, Maetani S, Morikawa N, Iwata H, Ikada Y. A newly developed three-layer agarose microcapsule for a promising biohybrid artificial pancreas: rat to mouse xenotransplantation. Cell Transpl 1996;5:S59–S63.

Tissue Engineering for Therapeutic Use 1.
Y. Ikada and Y. Yamaoka, editors.

Embryonic stem cell lines as a source for producing differentiated cells in vitro

Shin-Ichi Nishikawa
Department of Molecular Genetics, Faculty of Medicine, Kyoto University, Kyoto, Japan

Keywords: endothelial cell, ES cell, hemangioblast, PDGF receptor family.

Murine embryonic stem cell (ES) line is the only totipotent cell line that is presently available, while extensive attempts are being made to establish such lines from other species. Its usefulness has been proved in the last decades, as many mutant mice have been created from ES cells whose genomic genes were mutated by homologous recombination.

Because of its totipotency, the ES cell should have potential in the production of many types of mature cells which could be used as a source of cell supplement treatment. Its use in this direction, however, has been hampered by a number of obstacles. The most important obstacle is the difficulty to control the differentiation of ES cells into a given desired cell lineage. Indeed, while ES cell differentiation can be blocked by addition of leukemia inhibitory factor (LIF) in the culture, its removal induces uncontrolled differentiation into various cell lineages and the culture is used to hold a mixture of mature cells and unidentified cells in the intermediate stage of differentiation.

This problem could be solved in two ways. One, and probably the best way, is to use a combination of cytokines that can drive the ES cell differentiation to select a given pathway. However, we still need to understand more about the molecules controlling cell specification processes. The other way which is easier to achieve at this moment is to repeatedly select intermediate stages of a desired cell lineage from undesired ones until pure and fully differentiated cells are eventually produced in vitro. In this article, we will present our attempts along with the latter direction to produce vascular endothelial cells and blood cells from ES cells.

Differentiation pathway into hemangioblast, a putative precursors for hematopoietic cells and vascular endothelial cells

Among the many cell types constituting mammalian body, vascular endothelial cells and erythrocytes are those which appear most rapidly in the embryogenesis. In the murine embryo, both vascular endothelial cells and erythrocytes appeared

Address for correspondence: Dr Shin-Ichi Nishikawa, Department of Molecular Genetics, Kyoto University, Konoe-cho, Yoshida, Sakyo-ku, Kyoto 606-01, Japan.

in the yolk sac of embryo on 7.5 gestation, while most of the other cell types remain at the intermediate stage of differentiation. This suggests that the intermediate steps present in the differentiation process from totipotent embryonic ectoderm to hematopoietic and angiogenic cells may be less than those during differentiation of other cell types, depending on differentiation speed. In fact, only three distinct steps have been defined during this course, which are depicted in Fig. 1. From the fertilized egg to the embryonic ectoderm of the egg cylinder stages, each embryonic cell preserves multipotency. The first step of cell specification occurs at E6.5, when a structure called primitive streak is formed in the egg cylinder. Fate map study indicated that mesoderm and endoderm cells are generated only by passing through this structure [1]. With respect to the hematopoietic and vascular endothelial cell lineages, the cells residing in the most proximal part of primitive streak differentiate into proximal lateral mesoderm, which then differentiate into extraembryonic mesoderm and lateral plate mesoderm. Both can give rise to vascular endothelial cells. Moreover, recent studies strongly

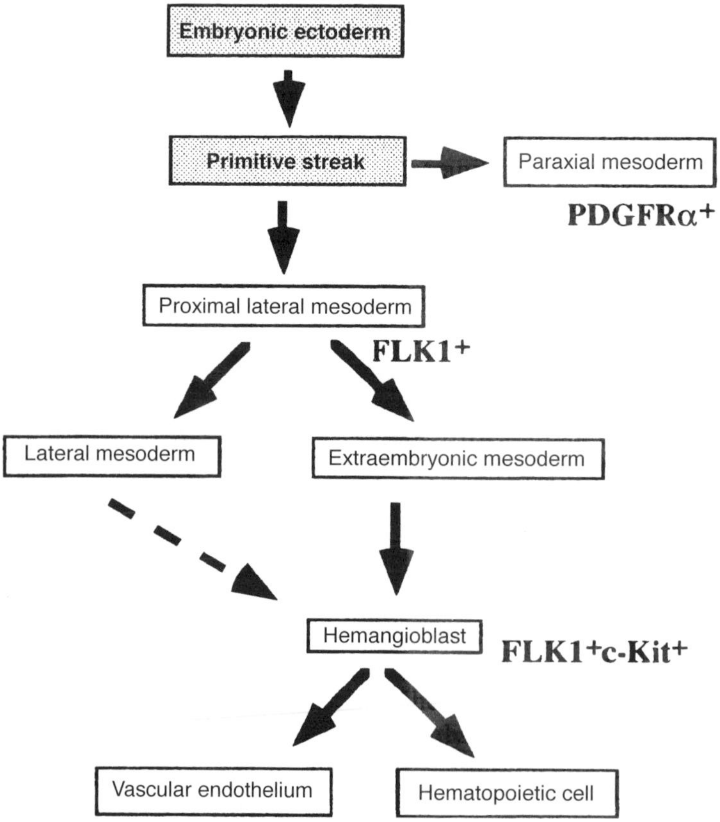

Fig. 1. Differentiation pathway of angio/hematopoietic cell lineages from totipotent embryonic ectoderm. Totipotent stages areindicated by dotted box. See text for details.

suggest that some progenies of the lateral plate mesoderm can give rise to hematopoietic cells [2]. Hence, two mesoderms are nearly the same with respect to their differentiation potency, although those migrating to the extraembryonic region differentiate most rapidly into mature erythrocytes and vascular endothelial cells through a stage called hemangioblast [3]. As our purpose is to obtain a culture system by which to generate only hematopoietic cells and vascular endothelial cells, the issue is whether or not each intermediate cell in this differentiation pathway is distinguished from other cell lineages and to eventually collect them at high purity for further differentiation culture.

Receptor tyrosine kinases of platelet-derived growth factor (PDGF)-receptor family (RTKP) are suitable surface markers to isolate early mesoderm cell lineages

With respect to hematopoietic and vascular cell lineages, the proximal lateral mesoderm is the first intermediate stage after the embryonic ectoderm. There were a number of molecules shown to be specifically expressed in this mesoderm, most of which were intracellular molecules and, therefore, difficult to use to stain living cells. Upon searching mesoderm specific molecules which could be potentially expressed on the cell surface, we found reports suggesting that RTKP are expressed specifically in mesoderm cells. To confirm these studies, we established monoclonal antibodies (mAb) to PDGFRα c-Kit, and Flk1, and tested if those mAbs could stain mesoderm cells specifically. Although c-Kit was expressed in various cell lineages besides mesoderm cells, PDGFRα and Flk1 were specifically expressed in the mesoderm cells of the gastrulating embryo. More interestingly, whole mount immunohistochemistry of embryos demonstrated that PDGFRα and Flk1 were expressed in the distinct sets of mesoderm cells, corresponding to paraxial and proximal lateral mesoderms respectively (Fig. 1). The fact that no RTKP positive cells were present in the primitive streak of the embryo indicated that only the mesoderm cells exfoliated from the primitive streak were specified by these mAbs.

It is important to note that our mAbs recognized the extracellular domain of RTKPs, so that basically they can be used for marking and sorting living cells expressing these markers. Indeed, the staining of living cells dissociated from the gastrulating embryo demonstrated that there were distinct cell populations expressing those RTKs in various combinations.

Other molecular markers defining earlier or later stages of differentiation

As mesoderm is an early intermediate stage in the course of differentiation from embryonic ectoderm to endothelial and hematopoetic cells, it is important to distinguish it from earlier and later stages by other surface markers. For this purpose, we used E-cadherin for the early stage, vascular endothelial cadherin (V-Cadherin) for vascular endothelial cell and CD45 and Ter119 for hematopoietic cell. E-cadherin, which was first identified by Takeichi's group in Kyoto University [4], has been shown to be expressed from eight-cell embryo to embryonic ectoderm but not in the mesoderm populations. On the other hand, V-Cadherin was shown to be expressed in the mature vascular endothelial cells and CD45, the common leukocyte antigen, has been used as a specific marker for hematopoietic cell lineage though not erythroid cell lineage.

By using these markers for staining gastrulating embryo, we confirmed that these surface markers can be utilized for defining the specified vascular endothelial cells and hematopoietic cells.

A culture to induce mesoderm subsets from ES cells

Basically, two methods have been employed to induce angiohematopoietic cell differentiation of ES cells in vitro; embryoid body culture and stromal cell dependent culture [5–7]. Removal of LIF is essential for inducing ES cell differentiation in both methods. One method first allows ES cells to form spherical epithelial body called embryoid body, in which many other cell lineages including hematopoietic cells are generated spontaneously. As this process appeared to follow the process of early embryogenesis, namely formation of egg cylinder and subsequent mesoderm generation, this method gained popularity. However, since all differentiation processes occur spontaneously inside the sphere enclosed in the epithelial sheet, the differentiation process is found to be difficult to observe and manipulate.

The other method is to use stromal cell lines to serve an appropriate microenvironment for the differentiation of ES cells into hematopoietic cells. A recent report of Nakano et al. demonstrated that a stromal cell line OP9 served as microenvironment to induce various hematopoietic cell lineages [7]. However, as expected, stromal-cell-based microenvironment is different from what is formed in the embryo, so that the culture itself is selective to restricted cell lineages.

Nevertheless, it is clear that any presently available culture conditions are selective, as we are unable to reproduce embryogenesis in vitro. Therefore, the extent to which these in vitro processes in the microenvironments: whether it is in the embryoid body or in the stromal-cell-dependent culture, can recapitulate the actual process within the embryo, should be evaluated by a set of markers for each intermediate stage. Thus, we used the surface markers described in the preceding section for determining which of the intermediate stages were able to be detected in the ES cell differentiation.

In addition, we have attempted to establish a culture in which ES cell differentiates into various mesoderm cells. After a number of preliminary experiments, we eventually reached a condition in which type IV collagen was used as matrix, but otherwise the condition was the same as that reported by Nakano et al. As depicted in Fig. 2, both PDGFRα^+ and FLK1$^+$ cells representing nascent mesoderm were detected in this culture as early as day 3 after removal of LIF. The populations have never been detected in the ES cell culture with LIF. More inter-

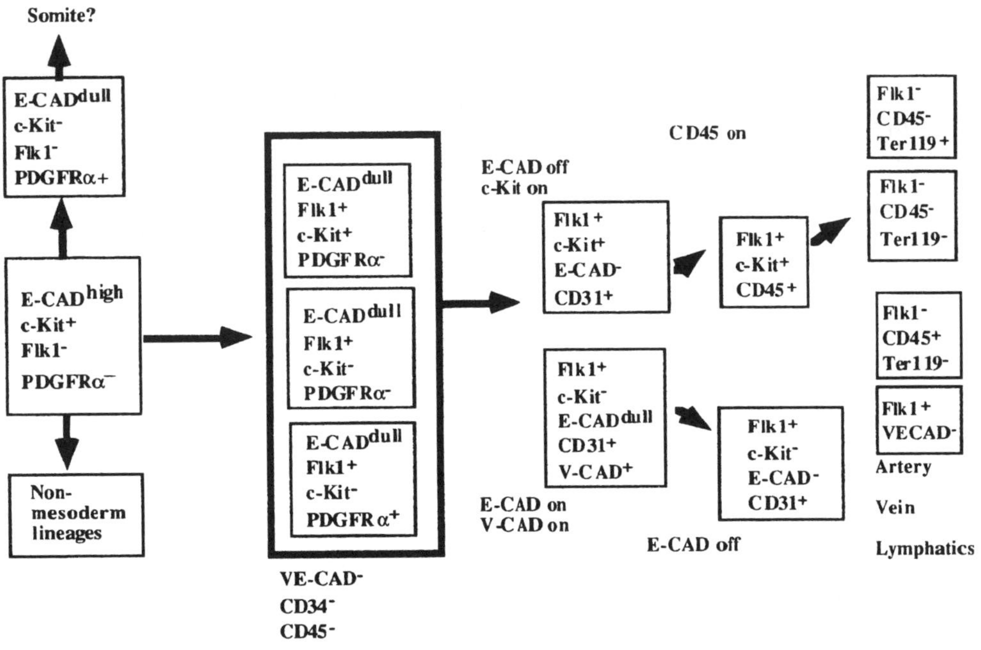

Fig. 2. Intermediate stages detectable during differentiation of angiohematopoietic cells.

estingly, nearly no V-Cadherin$^+$, CD45$^+$, or Ter119$^+$ cells were detected under our culture condition, even though they were incubated for up to 6 days. This indicated that our culture condition was permissive for mesoderm cells, but not for more differentiated cells.

Next, we sorted each subset of cells and recultured them on a stromal cell clone OP9 that has shown to support ES cell differentiation into hematopoietic cells. As expected from the embryonic development, only FLK1$^+$ cells are able to give rise to vascular endothelial cells and hematopoietic cells. Of note is that V-Cadherin$^+$ cells were generated as early as 1 day after the reculture of Flk1$^+$V-Cadherin$^-$ cells. Likewise, a substantial number of CD45$^+$ and Ter119$^+$ cells were detectable on day 2 of reculture. Thus, all necessary molecules for angiohematopoietic differentiation are expressed on OP9. Identification of such molecules would be the important issue for the future.

In addition to the mature vascular endothelial cells and hematopoietic cells, a number of intermediate stages from proximal-lateral mesoderm appeared to be identified. Those included are Flk1$^+$c-Kit$^+$V-Cadherin$^-$CD45$^+$ and Flk1$^+$c-Kit$^{-/dull}$V-Cadherin$^+$CD45$^-$ cells corresponding to the cells committed to hematopoietic cell and vascular endothelial cell respectively. In any case, we were thus able to select the intermediate stage that gives rise to angiohematopoietic cell-lineages and able to produce them specifically in vitro.

Prospect

Figure 2 summarizes our hypothesis derived from a round of cell surface staining and sorting during in vitro ES cell differentiation. At this moment, we are focusing on the very initial steps characterized by the expression of three RTKP, a part of which was presented in this symposium. Obviously, we need to extend our analysis from mesoderm to more mature intermediate stages listed in Fig. 2. If this figure represents the actual process, we will be able to complete the differentiation map, which will be useful to sort a more committed population. We also hope that this method will be extended to other cell lineages such as nerve cell lineage or primordial germ cells, which should be present in the population but which were not investigated in our study.

Acknowledgements

This study was supported by grants from the Japanese Ministry of Education, Science and Culture (No. 07CE2005, 07457085 and 06277103), a grant from the Ministry of Science and Technology (No. 13073-2125-14) and a grant from RIKEN.

References

1. Tam PP. Regionalization of mouse embryonic ectoderm: allocation of prospective ectodermal tissues during gastrulation. Development 1989;107:55–67.
2. Cumano A, Dieterlen-Lievre F, Godin I. Lymphoid potential, probed before circulation in mouse, is restricted caudal intraembryonic splanchnopleure. Cell 1996;86:907–916.
3. Pardnaud L, Yassine F, Dieterlen-Lievre F. Relationship between vasculogenesis, angiogenesis and hemopoiesis during avian ontogeny. Development 1989;100:339–349.
4. Takeichi M. Cadherins cell adhesion receptors as morphogenetic regulator. Science 1991;252: 1451–1455.
5. Doetchman TC, Eistter H, Katz M, Schmidt W, Kemler R. The in vitro development of blastocyst-derived embryonic stem cell lines: formation of visceral yolk sac, blood islands and myocardium. J Embryol Exp Morphol 1985;87:27–45.
6. Wiles MV, Keller G. Multiple hematopoietic lineage develop from embryonic stem cells in culture. Development 1991;111:259–267.
7. Nakano T, Kodama H, Honjo T. Generation of lymphohematopoietic cells from embryonic stem cells in culture. Science 1994;265:1098–1101.

Tissue Engineering for Therapeutic Use 1.
Y. Ikada and Y. Yamaoka, editors.

Prospects for the clinical use of xenogeneic organs and cells

Jeffrey L. Platt
Departments of Surgery, Pediatrics and Immunology, Duke University, Durham, North Carolina, USA

Keywords: antibody, antigen, compatibility, liver support, rejection, transplantation, xenotransplantation.

Introduction

At the beginning of the twentieth century, advances in surgery, such as the development of the vascular anastomosis, brought to the fore the possibility that an healthy organ, or tissue, might be transplanted as a treatment for disease. Indeed, studies by Carrel, Guthrie, Ullman and others established that the transplantation of the kidney, and perhaps of other organs, was feasible from a technical perspective [1–3]. What was not so certain then was what source of organs and tissues might be used to apply these techniques in human patients for the laws and customs of that time would not countenance the harvesting of tissue from human cadavers. Two potential solutions to this problem were envisioned. First, there was the possibility of using animals in lieu of humans as a source of transplants, that is xenotransplantation. Second, there was the possibility that human cells or tissue might be "grown", that new organs would be engineered. Despite the passage of nearly 90 years, neither of these approaches has gained routine clinical application; yet, today there is renewed excitement about the possibility that xenotransplantation and tissue engineering may be nearing the clinical arena and thus perhaps claiming the success which has for so long eluded these fields. This paper reviews what in 1997 appear to be some of the hurdles to the clinical application of xenotransplantation and how these hurdles can be addressed though the application of preclinical models.

The current interest in the clinical application of interspecies or xenotransplantation stems predominantly from a severe and worsening shortage of donors, which in the USA limits the number of transplant procedures to 5–50% of the number that might be performed if animals might be used as organ and tissue donors in lieu of humans [4]. This shortage of transplant donors is even more severe in certain other countries. Interest in xenotransplantation also stems from the possibility that the xenotransplant might be less susceptible than a human organ or tissue transplant to recurrence of viral or immunological disease. This

Address for correspondence: Dr Jeffrey L. Platt MD, Departments of Surgery, Pediatrics, and Immunology, Duke University Medical Center, Box 2605 DUMC, Durham, NC 27710, USA. Tel.: +1-919-681-3857. Fax: +1-919-681-7263. E-mail: platt001@mc.duke.edu possibility led to recent efforts to transplant the livers of baboons into patients with chronic hepatitis [5] and the bone marrow of a baboon in to a patient with HIV-1. The author has also proposed the possibility that the xenotransplant might be used advantageously as a vehicle to gain expression of genes of interest for the treatment of disease. Thus, by genetic engineering of a transplant donor, as has been carried out in the development of animals, the organs of which would resist rejection [6], one might deliver in a standard way genes to address genetic or metabolic disease.

A classification of xenografts and definition of terms

It is helpful to classify xenografts according to the manner in which the vascular supply is derived and the phylogenetic distance between the donor and the recipient (Table 1). Grafts consisting of free tissues, such as pancreatic islets or skin derive their vascular supply in part by the in-growth of recipient blood vessels. The survival and function of these grafts thus depends on the ability of the foreign microenvironment to support angiogenesis and differentiated function of the graft. For example, xenogeneic bone marrow may fail to engraft and undergo normal differentiation because the growth factors of the recipient are incompatible with the stem cells of the donor [7]. One important function of the microenvironment is the support of neovascularization. Thus, if vascularization of a free-tissue graft does not occur, the foreign tissue may die or, like corneal allografts and some xenografts, survive and evade severe humoral immune injury [8–10]. With the establishment of a vascular supply, free-tissue xenografts are subject to destruction by immune-mediated mechanisms, which as discussed below may differ qualitatively or quantitatively from the reaction observed in allografts. It is perhaps axiomatic that the more genetically disparate the graft and the host, the more difficult will be the biological and immunological hurdles to free-tissue transplantation. Still, the transplantation of pancreatic islets and central nervous system tissues has succeeded experimentally to the extent that clinical trials are now in progress [11–14].

Grafts consisting of intact organs are vascularized by surgical anastomosis of donor and recipient blood vessels. Organ xenografts provide their own micro-

Table 1. Classification of xenografts.

Type of xenograft (example)	Type of vascular supply	Microenvironment
Isolated cells (hepatocytes, bone marrow)	Neovascularization	Recipient
Free tissue (pancreatic islets, skin)	Neovascularization and anastomosis of donor and recipient vessels	Donor and recipient
Organ (kidney, heart)	Primary anastomosis of donor and recipient vessels	Donor

environment and thus are less dependent on the biological compatibility of the graft with the recipient. However, organ xenografts are confronted immediately by immune and inflammatory components of the recipient's blood and these components, especially naturally occurring antibodies and complement, may in some combinations of donor and recipient species cause immediate or hyperacute rejection of the graft. These species combinations are called "discordant". Species combinations in which hyperacute rejection of vascularized xenografts is not regularly seen are called "concordant". Until recently, hyperacute rejection was viewed as the major hurdle to transplantation of organs between species. With the recent discovery of the molecular basis for discordance of porcine-to-primate xenografts there has been increased hope that this barrier is in fact surmountable.

Selection of donors for xenotransplantation

In view of the preceding, it is intuitive that the most suitable donor of a xenograft is a species phylogenetically close to the recipient, in the case of humans such species are the higher, nonhuman primates. Indeed, 30 years ago, chimpanzees and baboons were used as donors for kidney transplants [15,16] and more recently baboons were used as donors for liver and bone marrow [5]. While the advantages of using nonhuman primates might seem overwhelming, there are some rather significant limitations. One important limitation is that the number of large primates that might be available to provide organs of suitable size is very small, far to small to meet the need for human organs. Another limitation is that primates might transmit infectious diseases which could spread to the general population [17]. Thus, while some, very limited, use of nonhuman primates has been proposed, most investigators favor the use of lower mammals, especially the pig, in lieu of primates as xenotransplant donors, despite the formidable biological hurdles involved in doing so. Not only is the pig available in large numbers and in sizes suitable for providing organs for transplantation into humans, as discussed below, but there is also much less need to be concerned about zoonotic agents. Use of the pig has the additional advantage that it allows the application of genetic engineering. These issues can be given only superficial consideration here and so the reader is referred to some recent reviews which address the subjects in greater detail [18—22].

The immunological hurdles to xenotransplantation

The major limitation to the clinical application of xenotransplantation is the immunological reaction of the recipient against the graft. This reaction can lead to the very rapid and violent rejection of the xenotransplant and in some cases may represent the most severe immune reactions known. Figure 1 shows a model for the immunological barriers to xenotransplantation of organs between disparate species.

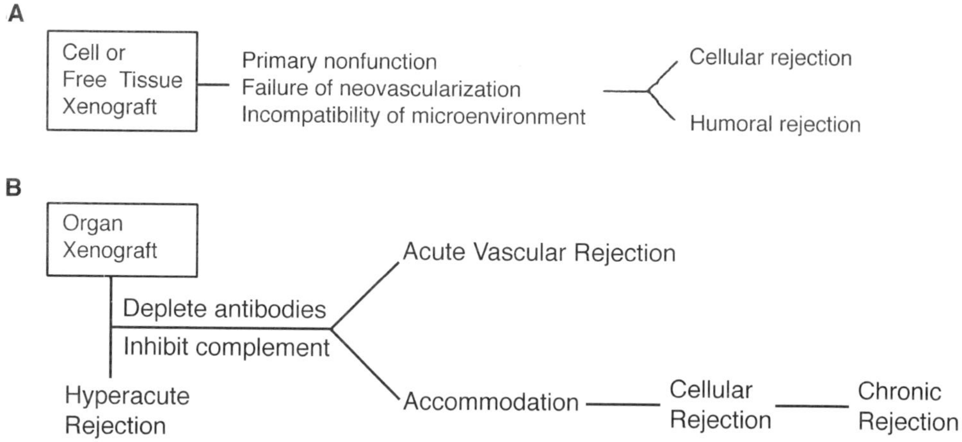

Fig. 1. The immunological response to xenotransplantation. The immune response to xenotransplantation can be classified according to whether the graft consists of isolated cells or free tissues, such as islets of Langerhans or of a primarily vascularized organ such as the kidney or heart. **A:** Cell and free-tissue grafts are subject to failure caused by primary nonfunction, failure of neovascularization or failure of the microenvironment to support the survival and function of the foreign tissue. If the free tissue or isolated cells engraft, they are then subject to cellular or humoral rejection. **B:** Vascularized organ grafts are subject to hyperacute and acute vascular rejection caused by the action of antidonor antibodies on donor endothelium. If hyperacute or acute vascular rejection are averted, the graft may undergo accommodation, a condition in which the graft appears to resist injury despite the return of antidonor antibodies to the circulation and the presence of an intact complement system. A vascularized organ graft may also be subject to cellular rejection and chronic rejection more or less like the corresponding types of rejection observed in allografts.

Hyperacute rejection

The first, and once considered the most daunting, barrier to xenotransplantation of porcine organs into humans is the susceptibility of such grafts to hyperacute rejection [23,24]. Hyperacute xenograft rejection is only seen in whole organ grafts and in humans and nonhuman primates it is initiated by the binding of xenoreactive natural antibodies to graft endothelium. The importance of xenoreactive antibodies in triggering hyperacute rejection of porcine xenografts by primates was suggested by the observation that a primate from which these antibodies were immunoabsorbed, leaving the complement system intact, does not hyperacutely reject a porcine heart [25,26]. Antibody binding activates the complement system which in turn mediates the tissue injury that ensues.

Xenoreactive antibodies and the antigens they recognize

The molecular basis for the rejection of porcine organs by humans and nonhuman primates has been elucidated. Human xenoreactive natural antibodies predominantly recognize Galα1-3Gal [27–29], a saccharide expressed on the cells of lower mammals and New World monkeys but not by Old World monkeys, apes or humans [30,31]. Antibody binding to this sugar expressed as a modification on endothelial cell components activates the complement by the classical pathway [25,32].

Complement

Activation of the complement system is an essential step in the pathogenesis of hyperacute rejection. Indeed, in every experimental model studied to date, inhibition of complement has effectively prevented hyperacute rejection [33]. The elements of the complement system which seem to be most important for the pathogenesis of hyperacute rejection are the terminal complement complexes (C5b6789). Animals deficient of complement components C5 or C6 do not hyperacutely reject disparate organ grafts [34—36]. However, the full complement cascade may not be needed, as hyperacute rejection may still occur if the complement is inhibited at the level of C8 [37].

Immunopathogenesis of hyperacute rejection

How the activation of the complement causes so dramatic a process as hyperacute rejection has been an important question. It was once thought that complement activation would simply "dissolve" or destroy blood vessels. While endothelial cell death may occur in some cases, the appearance of the early signs of hyperacute rejection in the face of intact endothelium suggests that other, non-cytotoxic, mechanisms must be important. One potential mechanism involves a change in endothelial cell shape leading to loss of vascular integrity mediated by terminal complement complexes [38]. This change can be brought about by the insertion of C5b67 complexes into endothelial cells and may thus explain how hyperacute rejection might proceed without the full complement cascade. Another mechanism may involve the loss of heparan sulfate from graft endothelial cells [39,40] (for discussion of these mechanisms, see [41]).

In addition to being the target of humoral reactions leading to complement activation, xenogeneic organ grafts are excessively susceptible to injury by the complement system of the recipient [23,42]. This susceptibility reflects the failure of the normal control mechanism to function adequately against the complement of other species. Thus, under physiological conditions, cells express complement inhibitory proteins such as decay-accelerating factor (DAF; CD55) and membrane cofactor protein (MCP; CD46), which inhibit C3 and C4, and CD59 (membrane inhibitor of reactive lysis), which inhibits C8 and C9. These proteins protect against inadvertent injury caused by activated complement components. These complement inhibitory proteins function poorly against heterologous complement; thus, the cells of a xenogeneic organ graft do not effectively regulate activation of the recipient's complement system. McCurry [43] and Byrne [44] recently demonstrated that expression of even low levels of human DAF and

CD59 by transgenic pigs protects the organs of such pigs from hyperacute rejection following transplantation into a nonhuman primate. As an exciting extension to this idea, White [45] announced the achievement of prolonged xenograft survival using pigs expressing DAF as donors for nonhuman primates.

Therapeutic approaches to hyperacute rejection

It has been apparent for some time that hyperacute rejection can be reliably prevented by inhibition of the complement system. Until recently, however, complement inhibition was accomplished by administration of agents such as cobra venom factor [46,47], which were thought to be too toxic for use in humans. Recent studies have revealed, however, that the complement can also be inhibited by agents such as gammaglobulin [48] and soluble CR1 [49], which have been used in human subjects for other complement-mediated processes. The major limitation to using complement inhibitors for the prevention of hyperacute rejection is that the inhibitors may compromise the ability to defend against infectious organisms, especially if they inhibit the complement at the level of C3. Thus, the development of transgenic animals expressing human complement regulatory proteins as components of cell membranes of the graft would seem to allow complement inhibition to be focused at the cell surface, leaving the complement cascade intact for host defense. Another approach which prevents hyperacute rejection is the removal of xenoreactive antibodies from the recipient [26,50–53]. The identification of the antigen recognized by xenoreactive antibodies now makes the specific removal of these antibodies feasible and allows for the development of strategies for the modification of cell surfaces to decrease antibody binding. This aspect of therapy is discussed below.

Acute vascular xenograft rejection

If hyperacute rejection is averted, a xenograft becomes subject to a type of rejection which we originally described and named acute vascular xenograft rejection [54]. Sometimes called delayed xenograft rejection, acute vascular xenograft rejection begins characteristically within 24 h in highly disparate xenografts or within 2–3 days of the transplantation of concordant xenografts. Given the ability to prevent hyperacute rejection, acute vascular xenograft rejection is now seen as the major immunological hurdle to xenotransplantation.

Role of xenoreactive antibodies in acute vascular rejection

Although the cause of acute vascular xenograft rejection is a subject of current controversy, some important clues have recently emerged. There is accumulating evidence that acute vascular xenograft rejection is caused at least in part by the unremitting interaction of xenoreactive antibodies with donor blood vessels. Coinciding with what would be expected to be the onset of acute vascular xenograft rejection, a striking increase in the levels of xenoreactive antibodies occurs in human subjects whose blood was perfused through porcine livers as a treatment for hepatic failure [55]. In immunosuppressed individuals, these antibodies are predominantly directed against Galα1-3Gal. Depletion of xenoreactive antibodies from a xenograft recipient delays or prevents acute vascular rejection [56]. Further evidence for the importance of antidonor antibodies in acute vascular rejection can be taken from experimental studies in which the transfer of immune serum leads to the rapid onset of vascular rejection [57].

Immunopathogenesis of acute vascular rejection

Binding of xenoreactive antibodies to donor blood vessels may eventuate in the development of acute vascular rejection through one or more of several pathways. First, xenoreactive antibodies specific for Galα1-3Gal may lead to perturbation of endothelial cell functions or activation of endothelial cells. This concept emerges from the recent finding that Gala1-3Gal is expressed as a substitution on endothelial cell membrane integrins [32], the known functions of which include the maintenance of endothelial integrity and transmembrane signaling [58–60]. Second, antibody binding may lead to activation of small amounts of the complement on endothelial cells in turn leading to the activation of the cells by secretion and autocrine functioning of IL1α [61]. This latter mechanism may be important not only in acute vascular rejection, but also in other types of vasculitis caused by humoral immunity.

Although acute vascular rejection is generally described as a phenomenon of organ grafts, it seems likely that vascular rejection might also occur in free-tissue grafts. Consistent with that are reports of antibody-mediated rejection of skin xenografts in rodent models [62,63].

Therapeutic approaches to acute vascular xenograft rejection

Disrupting the interaction of xenoreactive antibody and antigen is the mainstay of therapy for acute vascular xenograft rejection. The physical removal of xenoreactive antibodies has already been mentioned [56]. This approach should be combined with immunosuppressive therapy, such as administration of cytotoxic agents, that optimally inhibit antibody production. Another strategy involves decreasing the expression of the antigenic target of xenoreactive antibodies. One approach to decreasing antigen expression is derived from the studies of Geller [64] and Alvarado [65] who showed that antigen expression varies among pigs and that this variation has a genetic basis. Thus, donors could be selected or bred. Recent studies by Sandrin [66] and Sharma [67] suggest that expression of α1,2-fucosyltransferase in transgenic animals diverts carbohydrate synthesis from production of Galα1-3Gal to production of the H antigen which is not targeted by natural antibodies.

Elicited cellular and humoral immune barriers to xenotransplantation

Organ xenografts and free-tissue xenografts are subject to rejection by elicited cellular and humoral immune responses. Recent successes in overcoming the initial immunological hurdles to organ xenotransplantation and success in transplanting free tissues across wide phylogenetic barriers has sparked interest in these elicited immune responses as the next major hurdle to xenotransplantation. The potential importance of these responses is suggested by the significant barrier to achieving long-term survival of concordant organ xenografts.

There are at least two pre-eminent questions regarding the cellular immune responses to xenotransplantation. The first question is whether the cellular immune response to a xenograft would differ quantitatively or qualitatively from the cellular immune response to an allograft. The second question is whether the xenoimmune response can be addressed by the same therapeutic measures as the alloimmune response.

It was once thought on theoretical grounds that xenoimmune responses might be less intense than alloimmune responses. The basis for this concept was the observation that mixed leukocyte cultures, which are thought to model the events leading to transplantation immunity, give rise to vigorous responses to allogeneic cells but very much dampened or undetectable responses to xenogeneic cells [68—71]. Allogeneic cells are recognized in such cultures predominantly through the direct pathway, the host's T cells recognizing donor MHC molecules as native structures on donor cells, and as a result the alloimmune response is very intense and rapid (there being a large percentage of host T cells able to participate in such a response). In contrast, xenogeneic cells were thought to be recognized predominately through the indirect pathway, the host's T cells being specific for peptides of xenogeneic origin complexed with host MHC on host antigen presenting cells. Since indirect recognition would lead to activation of a smaller fraction of the T cell repertoire, the response arising in this way would be less intense and less rapid that the allogeneic response. Recent studies have shown, however, that the major defect in the xenogeneic responses is in the interactions between pairs of cell adhesion and costimulatory molecules, that T cells can recognize xenogeneic cells directly, and that under the proper conditions xenoimmune responses arising in vitro may be as vigorous as alloimmune responses [70,72,73].

Some other aspects of the cellular immune response to xenogeneic cells remain to be elucidated. To which extent the cellular response will be focused against MHC as it is in allografts vs. other proteins is uncertain. Also uncertain is whether the response will be subject to the kinds of regulation that impact on alloimmune responses. Finally, there is the likelihood that humoral responses will impact on the cellular immune response [74]. Whether as a result of these various factors cellular xenoimmune responses arising in vivo will also be vigorous remains to be formally tested.

In addition to the response posed by T cells there has been interest in the possibility that natural killer (NK) cells may contribute to graft injury. NK cells have receptors which recognize autologous cells and upon doing so down regulate natural killer activity [75]. The failure of these receptors to recognize allogeneic bone marrow cells and thus to inhibit natural killer activity is thought to be responsible for hybrid resistance following bone marrow transplantation [76]. Presumably, NK cells would not recognize a xenogeneic cell and thus effector activity would be unrestrained. Consistent with this idea we and others have observed that human NK cells mediate killing and various pathophysiologic changes in heterologous endothelial cells [77,78]. The extent to which NK cells might cause similar changes in vivo remains to be determined, however. Of interest is evidence that the function of NK cells against heterologous targets is amplified by xenoreactive antibodies [79].

Another facet of the elicited immune response is the potential development of a variety of antidonor antibodies. Already discussed is the observation that under some conditions these antibodies may be restricted in their specificity to Galα1-3Gal. Still, it seems likely that antibodies against other donor antigens, particularly polypeptides, are likely to arise and pose an additional hurdle, perhaps in the form of acute or chronic vascular rejection. This question is clearly an important one for future investigation.

Therapeutic considerations

Whether cellular immune responses against a xenograft will be subject to treatment with conventional immunosuppressive agents is uncertain. There is concern that to the extent that natural killer cells contribute to the cellular reactions, these reactions may be less amenable to conventional immunosuppressive agents. Studies on the outcome of concordant xenografts suggest that the levels of immunosuppression and the regimens needed to bring about long-term graft survival may be much more severe that the corresponding levels and regimens for allotransplantation [80,81]. Conversely, it is not unlikely that some of this difference reflects the need to suppress humoral responses and there is limited evidence that cellular responses may be addressed with well-established approaches such as pulse therapy with methylprednisolone [82]. Furthermore, for some types of transplants, such as that of neural tissue and hepatocytes, there is the possibility of placing the foreign cells in an anatomical location which is relatively protected from immunological attack. Indeed, once a tissue of cellular transplant is revascularized by the host, the foreign cells are often found to be buffered from cellular immune injury. Of course, such protection from immune injury is not presently possible for an organ xenograft and therefore, other approaches are under active consideration. One approach to dealing with both humoral and cellular responses may be to induce tolerance to the xenogeneic donor. Among the strategies currently proposed is the induction of mixed hematopoietic chimerism [19]. Mixed chimerism may induce humoral tolerance by direct interaction between donor cells and recipient B cells or by inhibiting cellular immune responses through central mechanisms. There is also the possibility that the opportunity offered by xenotransplantation to apply genetic engineering will allow the introduction of genes which might help to decrease immune responses to the foreign tissue. For instance, some research is being carried out to explore the usefulness of expressing Fas ligand (CD95) on foreign tissues in the hope that T cells recognizing those tissues would undergo apoptosis and thus clonal deletion [83].

Other biological hurdles to xenotransplantation

Biological hurdles other than immunity may be important in xenotransplantation. The most fundamental hurdle may be the extent to which the transplanted organ or tissue can provide an adequate physiologic replacement. There is at least some experimental evidence suggesting that the porcine heart, kidney and lung would seem to function at a level which would support the life of a primate. Still unknown is whether subtle defects or limits in xenograft function will exist that would impose a burden on the recipient. On the other hand, there would seem to be good reason to question whether, or to which extent, a porcine liver xenograft would be able to replace the vital functions of the primate liver [84]. Chari recently showed that perfusion of a porcine liver ex vivo could correct the most life threatening of complications of fulminant hepatic failure [85], however, functional replacement of the primate liver by a porcine liver has been achieved for a period of days in only a few cases [86].

Another biological hurdle to xenotransplantation involves the potential incompatibility of the graft with the host cells or with the complex systems or cascades in the host blood. Already discussed is the possibility that the microenvironment of the donor tissue may fail to support angiogenesis and the severe problem posed by the incompatibility of porcine complement regulatory proteins with the human complement system. Another "incompatibility" was recently discovered by Lawson et al. who found that porcine thrombomodulin does not function effectively with human thrombin to activate protein C [87,88]. This defect could contribute to the procoagulant "posture" of a porcine xenograft. Whether other critical incompatibilities exist remains to be tested. One important consideration, however, is that while in vitro studies may reveal potentially important, incompatibilities, some of these defects may actually benefit the graft and some may be balanced by other, compensatory, defects. Thus, despite all of the system which might function suboptimally, it has been possible to achieve enduring survival of highly disparate xenografts under some circumstances.

Zoonosis as a hurdle to xenotransplantation

Transplantation of cells tissues or organs from one individual to another engenders a risk of transmission of infectious disease from the donor to the recipient. This risk is of especial concern in allotransplantation where the transmission of cytomegalovirus or Ebstein-Barr virus occurs frequently and can be associated with significant medical complications. The possibility that a xenograft might transmit an infectious disease to the recipient has been a matter of concern and a subject of recent reviews [17,89]. Several aspects of this issue warrant discussion here as they pose potential hurdles to xenotransplantation.

While the actual risk of transmission of disease through a xenograft is impossible to predict at present, there is little question that risk is lower than the corresponding risk for clinical allografts. The lesser the risk of transmitting infectious diseases reflects in part the limited cross-species transmissibility of many viruses and in part the fact that in contrast to allotransplantation in xenotransplantation the donor can be screened for infectious agents and infected donors excluded where desired. Still, there is a finite risk that xenotransplantation will be the source of zoonosis. Thus, the clinician must weigh the risk of disabling or fatal complications from zoonosis against the corresponding risks of the patient's underlying disease. To the extent that xenotransplantation is a life-saving procedure, there may be a willingness to countenance a risk of disease transmission.

Beyond the potential risk for the patient, zoonotic disease is believed to pose a special risk for society. Thus, there is concern that "infectious" agents conveyed by a xenograft to an immunosuppressed patient might spread to the human population, initiating an epidemic. To this concern is added the possibility that the enduring presence of a xenogeneic transplant could allow the recombination of genetic elements to yield an infectious agent where none existed before. For example, a retrovirus endogenous to pig might recombine with human viruses to generate a novel pathogen. While such a scenario has not been documented in the limited number of cases in which clinical xenografts were performed, there have been two epidemics thought to have been initiated as zoonosis: HIV and swine influenza, which are commonly mentioned examples. It is the risk of such an outbreak that has restrained, to a certain extent, efforts to develop clinical programs in xenotransplantation and thus consideration of this risk now constitutes a major barrier to xenotransplantation.

To what extent should the consideration of zoonosis be allowed to restrain efforts to develop clinical xenotransplantation? Given the extent to which this question is one of public policy, it is beyond the scope of this review. However, there are certain factors which should be considered in this debate. One important factor is that there are no known agents in pigs which would pose a unique risk to society and which cannot be excluded from potential donors. For example, an endogenous porcine retrovirus isolated at various times in the past from a porcine cell line and recently shown to be able to infect and become transcriptionally active in human cell lines [90] has never been found to infect normal human cells or to cause any disease. Moreover, any agent which could be transmitted by a xenograft is likely (perhaps, more likely) to enter the human population through other means, such as transmission on farms, meat packing plants or in the home. Thus, the examples of zoonosis commonly cited – HIV and swine influenza – entered the human population by means other than xenotransplantation. Indeed, there is no known agent known which could only be transmitted by a xenograft and by no other means and, if such an agent existed, it would probably not spread easily from one human to another. Thus, weighing the potential benefits of xenotransplantation over the potential risks, it is not unreasonable to think that xenotransplantation is more likely to offer a therapeutic solution for the complications of a future zoonotic epidemic than it is to be the cause of such an epidemic.

Summary and prospects for clinical application of xenotransplantation

At the time that this article was prepared, in early 1997, there was increasing enthusiasm about the prospects for clinical xenotransplantation. In part this enthusiasm springs from the urgent need for donor organs and a sense that clinicians, as well as the public are willing to consider xenotransplantation as a rational approach to that problem. However, some of the enthusiasm also springs from new insights into the cellular and molecular mechanisms of xenograft rejection that have come to light. One reason for the renewed consideration given to xenotransplantation is the evidence that as molecular hurdles are discovered they can be overcome by genetic engineering of donor animals. This approach has only begun to be tested in large animals but already the results are encouraging. The important advantage of this approach over systemic therapy is that it focuses on the organ transplant, allowing the immune systems of the recipient to maintain a defense against invasive organisms. While it is not possible to predict with certainty when organ xenotransplantation will enter the clinical arena, it does seem likely that recent advances will be progressively tested and applied. For instance, the author and colleagues have already developed a clinical program for the use of organs from transgenic pigs as ex vivo devices for the treatment of fulminant hepatic failure. Others are using xenogeneic cells and tissues for the treatment of diabetes, and neurological and liver disease. As the immune hurdles yield to new therapeutic approaches, a next step will be temporary transplants. In addition, one can be confident that the eventual conduct of permanent transplants will benefit from insights gained in these early trials.

Acknowledgements

The work from the author's laboratory was supported by grants from the National Institutes of Health (HL46810, HL52297, HL50985 and DK38108) and from Nextran. This article was adapted from Platt JL, Am J Med Sci (In Press).

References

1. Carrel A. Transplantation in mass of the kidneys. J Exp Med 1908;10:98–140.
2. Guthrie CC. Blood-Vessel Surgery and its Applications. New York: Longmans, Green & Co., 1912.

3. Ullman E. Tissue and organ transplantation. Ann Surg 1914;60:195–219.
4. Evans RW, Orians CE, Ascher NL. The potential supply of organ donors: an assessment of the efficiency of organ procurement efforts in the United States. JAMA 1992;267:239–246.
5. Starzl TE, Fung J, Tzakis A, Todo S, Demetris AJ, Marino IR, Doyle H, Zeevi A, Warty V, Michaels M, Kusne S, Rudert WA, Trucco M. Baboon-to-human liver transplantation. Lancet 1993;341:65–71.
6. Platt JL, Logan JS. The generation and use of transgenic animals for xenotransplantation. In: Houdebine LM (ed) Transgenic Animals: Generation and Use. Amsterdam: Harwood Academic Publishers bv, 1997;(In press).
7. Gritsch HA, Glaser RM, Emery DW, Lee LA, Smith CV, Sablinski T, Arn JS, Sachs DH, Sykes M. The importance of nonimmune factors in reconstitution by discordant xenogeneic hematopoietic cells. Transplantation 1994;57:906–917.
8. Ben-Hur N, Solowey AC, Rapaport FT. The xenograft rejection phenomenon. II. Response of the guinea pig to mouse, rat and rabbit skin xenografts. Isr J Med Sci 1969;5:322–331.
9. Ben-Hur N, Solowey AC, Rapaport FT. The xenograft rejection phenomenon. I. Response of the mouse to rabbit, guinea pig and rat skin xenografts. Isr J Med Sci 1969;5:1–12.
10. Gunsalus JR, Brady DA, Coulter SM, Gray BM, Edge A. Reduction of serum cholesterol in watanabe rabbits by xenogeneic hepatocellular transplantation. Nature Med 1997;3:48–53.
11. Sun YL, Xiaojun M, Zhou D, Vacek I, Sun AM. Normalization of diabetes in spontaneously diabetic cynomolgus monkeys by xenografts of microencapsulated porcine islets without immunosuppression. J Clin Invest 1996;98:1417–1422.
12. Ricordi C, Lacy PE, Sterbenz K, Davie JM. Low-temperature culture of human islets or in vivo treatment with L3T4 antibody produces a marked prolongation of islet human-to-mouse xenograft survival. Proc Natl Acad Sci USA 1987;84:8080–8084.
13. Archer DR, Cuddon PA, Lipsitz D, Duncan ID. Myelination of the canine central nervous system by glial cell transplantation: a model for repair of human myelin disease. Nature Med 1997;3:54–59.
14. Isacson O, Deacon TW, Pakzaban P, Galpern WR, Dinsmore J, Burns LH. Transplanted xenogeneic neural cells in neurodegenerative disease models exhibit remarkable axonal target specificity and distinct growth patterns of glial and axonal fibers. Nature Med 1995;1:1189–1194.
15. Reemtsma K, McCracken BH, Schlegel JU, Pearl MA, Pearce CW, De Witt CW, Smith PE, Hewitt RL, Flinner RL, Creech O. Renal heterotransplantation in man. Ann Surg 1964;160: 384–410.
16. Starzl TE, Marchioro TL, Peters GN, Kirkpatrick CH, Wilson WEC, Porter KA, Rifkind D, Ogden DA, Hitchcock CR, Waddell WR. Renal heterotransplantation from baboon to man: experience with six cases. Transplantation 1964;2:752–776.
17. Chapman LE, Folks TM, Salomon DR, Patterson AP, Eggerman TE, Noguchi PD. Xenotransplantation and xenogeneic infections. N Engl J Med 1995;333:1498–1501.
18. Cooper DKC, Ye Y, Rolf LL Jr, Zuhdi N. The pig as potential organ donor for man. In: Cooper DKC, Kemp E, Reemtsma K, White DJG (eds) Xenotransplantation. The Transplantation of Organs and Tissues Between Species. New York: Springer-Verlag, 1991;481–500.
19. Sachs DH, Sablinski T. Tolerance across discordant xenogeneic barriers. Xenotransplantation 1995;2:234–239.
20. Platt JL. The immunological barriers to xenotransplantation. Crit Rev Immunol 1996;16: 331–358.
21. Parker W, Saadi S, Lin SS, Holzknecht ZE, Bustos M, Platt JL. Transplantation of discordant xenografts: a challenge revisited. Immunol Today 1996;17:373–378.
22. Platt JL. Xenotransplantation: recent progress and current perspective. Curr Opin Imm 1996;8: 721–728.
23. Platt JL, Vercellotti GM, Dalmasso AP, Matas AJ, Bolman RM, Najarian JS, Bach FH. Transplantation of discordant xenografts: a review of progress. Immunol Today 1990;11:450–456.
24. Platt JL. Hyperacute Xenograft Rejection. New York: Springer-Verlag, 1995.

25. Platt JL, Fischel RJ, Matas AJ, Reif SA, Bolman RM, Bach FH. Immunopathology of hyperacute xenograft rejection in a swine-to-primate model. Transplantation 1991;52:214–220.
26. Lin SS, Kooyman DL, Daniels LJ, Daggett CW, Parker W, Lawson JH, Hoopes CW, Gollotto C, Li L, Birch P, Davis RD, Diamond LE, Logan JS, Platt JL. The role of natural anti Galα1-3Gal antibodies in hyperacute rejection of pig-to-baboon cardiac xenotransplants. Transpl Immunol 1997;5:212–218.
27. Good AH, Cooper DKC, Malcolm AJ, Ippolito RM, Koren E, Neethling FA, Ye Y, Zuhdi N, Lamontagne LR. Identification of carbohydrate structures that bind human antiporcine antibodies: implications for discordant xenografting in humans. Transpl Proc 1992;24:559–562.
28. Sandrin MS, Vaughan HA, Dabkowski PL, McKenzie IFC. Anti-pig IgM antibodies in human serum react predominantly with Galα(1,3)Gal epitopes. Proc Natl Acad Sci USA 1993;90: 11391–11395.
29. Collins BH, Parker W, Platt JL. Characterization of porcine endothelial cell determinants recognized by human natural antibodies. Xenotransplantation 1994;1:36–46.
30. Galili U, Macher BA, Buehler J, Shohet SB. Human natural anti-α-galactosyl IgG: the specific recognition of α(1-3)-linked glactose residues. J Exp Med 1985;162:573–582.
31. Galili U, Clark MR, Shohet SB, Buehler J, Macher BA. Evolutionary relationship between the natural anti-Gal antibody and the Gal α1-3Gal epitope in primates. Proc Natl Acad Sci USA 1987;84:1369–1373.
32. Holzknecht ZE, Platt JL. Identification of porcine endothelial cell membrane antigens recognized by human xenoreactive antibodies. J Immunol 1995;154:4565–4575.
33. Auchincloss Jr H. Xenogeneic transplantation. Transplantation 1988;46:1–20.
34. Chartrand C, O'Regan S, Robitaille P, Pinto-Blonde M. Delayed rejection of cardiac xenografts in C6-deficient rabbits. J Immunol 1979;38:245.
35. Brauer RB, Baldwin III WM, Daha MR, Pruitt SK, Sanfilippo F. Use of C6-deficient rats to evaluate the mechanism of hyperacute rejection of discordant cardiac xenografts. J Immunol 1993;151:7240–7248.
36. Kroshus TJ, Rollins SA, Dalmasso AP, Elliott EA, Matis LA, Squinto SP, Bolman RM. Complement inhibition with an anti-C5 monoclonal antibody prevents acute cardiac tissue injury in an ex vivo model of pig-to-human xenotransplantation. Transplantation 1995;60:1194–1202.
37. Diamond LE, McCurry KR, Oldham ER, McClellan SB, Martin MJ, Platt JL, Logan JS. Characterization of transgenic pigs expressing functionally active human CD59 on cardiac endothelium. Transplantation 1996;61:1241–1249.
38. Saadi S, Platt JL. Transient perturbation of endothelial integrity induced by antibodies and complement. J Exp Med 1995;181:21–31.
39. Platt JL, Vercellotti GM, Lindman BJ, Oegema TR Jr, Bach FH, Dalmasso AP. Release of heparan sulfate from endothelial cells: implications for pathogenesis of hyperacute rejection. J Exp Med 1990;171:1363–1368.
40. Ihrcke NS, Platt JL. Shedding of heparan sulfate proteoglycan by stimulated endothelial cells: evidence for proteolysis of cell surface molecules. J Cell Physiol 1996;168:625–637.
41. Saadi S, Ihrcke NS, Platt JL. Pathophysiology of xenograft rejection. In: Lieberman R, Morris R (eds) Principles of Immunomodulatory Drug Development in Transplantation and Autoimmunity. New York, NY: Raven Press, 1996;31–45.
42. Dalmasso AP, Vercellotti GM, Platt JL, Bach FH. Inhibition of complement-mediated endothelial cell cytotoxicity by decay accelerating factor: potential for prevention of xenograft hyperacute rejection. Transplantation 1991;52:530–533.
43. McCurry KR, Kooyman DL, Alvarado CG, Cotterell AH, Martin MJ, Logan JS, Platt JL. Human complement regulatory proteins protect swine-to-primate cardiac xenografts from humoral injury. Nature Med 1995;1:423–427.
44. Byrne GW, McCurry KR, Martin MJ, McClellan SM, Platt JL, Logan JS. Transgenic pigs expressing human CD59 and decay-accelerating factor produce an intrinsic barrier to complement-mediated damage. Transplantation 1997;63:149–155.

45. Thompson C. Humanised pigs hearts boost xenotransplantation. Lancet 1995;346:766.
46. Gewurz H, Clark DS, Cooper MD, Varco RL, Good RA. Effect of cobra venom-induced inhibition of complement activity on allograft and xenograft rejection reactions. Transplantation 1967;5:1296–1303.
47. Leventhal JR, Dalmasso AP, Cromwell JW, Platt JL, Manivel CJ, Bolman RM, Matas AJ. Prolongation of cardiac xenograft survival by depletion of complement. Transplantation 1993;55: 857–866.
48. Magee JC, Collins BH, Harland RC, Lindman BJ, Bollinger RR, Frank MM, Platt JL. Immunoglobulin prevents complement mediated hyperacute rejection in swine-to-primate xenotransplantation. J Clin Invest 1995;96:2404–2412.
49. Pruitt SK, Kirk AD, Bollinger RR, Marsh Jr HC, Collins BH, Levin JL, Mault JR, Heinle JS, Ibrahim S, Rudolph AR, Baldwin III WM, Sanfilippo F. The effect of soluble complement receptor type 1 on hyperacute rejection of porcine xenografts. Transplantation 1994;57: 363–370.
50. Perper RJ, Najarian JS. Experimental renal heterotransplantation. I. In widely divergent species. Transplantation 1966;4:377–388.
51. Giles GR, Boehmig HJ, Lilly J, Amemiya H, Takagi H, Coburg AJ, Hathaway WE, Wilson CB, Dixon FJ, Starzl TE. Mechanism and modification of rejection of heterografts between divergent species. Transpl Proc 1970;2:522–537.
52. Cooper DKC, Human PA, Lexer G, Rose AG, Rees J, Keraan M, Du Toit E. Effects of cyclosporine and antibody adsorption on pig cardiac xenograft survival in the baboon. J Heart Transplant 1988;7:238–246.
53. Fischel RJ, Matas AJ, Platt JL, Perry E, Noreen H, Shumway SJ, Bolman RM. Cardiac xenografting in the pig to rhesus model; manipulation of antiendothelial antibody prolongs survival. J Heart Lung Transplant 1992;11:965–974.
54. Leventhal JR, Matas AJ, Sun LH, Reif S, Bolman RM III, Dalmasso AP, Platt JL. The immunopathology of cardiac xenograft rejection in the guinea pig to rat model. Transplantation 1993;56:1–8.
55. Cotterell AH, Collins BH, Parker W, Harland RC, Platt JL. The humoral immune response in humans following cross-perfusion of porcine organs. Transplantation 1995;60:861–868.
56. Lin SS, Weidner BC, Byrne GW, Diamond LE, Lawson JH, Hoopes CW, Daniels LJ, Daggett CW, Parker W, Harland RC, Davis RD, Bollinger RR, Logan JS, Platt JL. The role of antibodies in acute vascular rejection of pig-to-baboon cardiac transplants. J Clin Invest 1998;101:1745–1756.
57. Perper RJ, Najarian JS. Experimental renal heterotransplantation. III. Passive transfer of transplantation immunity. Transplantation 1967;5:514–533.
58. Lampugnani MG, Resnati M, Dejana E, Marchisio PC. The role of integrins in the maintenance of endothelial monolayer integrity. J Cell Biol 1991;112:479–490.
59. Hynes RO. Integrins: versatility, modulation, and signaling in cell adhesion. Cell 1992;69: 11–25.
60. Luscinskas FW, Lawler J. Integrins as dynamic regulators of vascular function. FASEB J 1994; 8:929–938.
61. Saadi S, Holzkhecht RA, Patte CP, Stern DM, Platt JL. Complement-mediated regulation of tissue factor activity in endothelium. J Exp Med 1995;182:1807–1814.
62. Winn HJ, Baldamus CA, Jooste SV, Russell PS. Acute destruction by humoral antibody of rat skin grafted to mice: the role of complement and polymorphonuclear leukocytes. J Exp Med 1973;137:893–910.
63. Baldamus CA, McKenzie IF, Winn HJ, Russell PS. Acute destruction by humoral antibody of rat skin grafted to mice. J Immunol 1973;110:1532–1541.
64. Geller RL, Rubinstein P, Platt JL. Variation in expression of porcine xenogeneic antigens. Transplantation 1994;58:272–277.
65. Alvarado CG, Cotterell AH, McCurry KR, Collins BH, Magee JC, Berthold J, Logan JS, Platt

JL. Variation in the level of xenoantigen expression in porcine organs. Transplantation 1995;59: 1589–1596.

66. Sandrin MS, Fodor WL, Mouhtouris E, Osman N, Cohney S, Rollins SA, Guilmette ER, Setter E, Squinto SP, McKenzie I. Enzymatic remodelling of the carbohydrate surface of a xenogeneic cell substantially reduces human antibody binding and complement-mediated cytolysis. Nature Med 1995;1:1261–1267.
67. Sharma A, Okabe JF, Birch P, Platt JL, Logan JS. Reduction in the level of gal (α1,3) gal in transgenic mice and pigs by the expression of an α(1,2) fucosyltransferase. Proc Natl Acad Sci USA 1996;93:7190–7195.
68. Lindahl KF, Bach FH. Genetic and cellular aspects of xenogeneic mixed leukocyte culture reaction. J Exp Med 1976;144:305–318.
69. Alter BJ, Bach FH. Cellular basis of the proliferative response of human T cells to mouse xenoantigens. J Exp Med 1990;191:333–338.
70. Kirk AD, Li RA, Kinch MS, Abernethy KA, Doyle C, Bollinger RR. The human antiporcine cellular repertoire. In vitro studies of acquired and innate cellular responsiveness. Transplantation 1993;55:924.
71. Moses RD, Winn HJ, Auchincloss Jr H. Multiple defects in cell surface molecule interactions across species differences are responsible for diminished xenogeneic T cell responses. Transplantation 1992;53:203–209.
72. Murray AG, Khodadoust MM, Pober JS, Bothwell ALM. Porcine aortic endothelial cells activate human T cells: direct presentation of MHC antigens and costimulation by ligands for human CD2 and CD28. Immunity 1994;1:57–63.
73. Yamada K, Sachs DH, Der Simonian H. Human antiporcine xenogeneic T cell response. Evidence for allelic specificity of mixed leukocyte reaction and for both direct and indirect pathways of recognition. J Immunol 1995;155:5249–5256.
74. Ihrcke NS, Wrenshall LE, Lindman BJ, Platt JL. Role of heparan sulfate in immune system-blood vessel interactions. Immunol Today 1993;14:500–505.
75. Lanier LL, Phillips JH. Inhibitory MHC class I receptors on NK cells and T cells. Immunol Today 1996;17:86–91.
76. Kumar V, George T, Yu YYL, Liu J, Bennett M. Role of murine NK cells and their receptors in hybrid resistance. Curr Opin Imm 1997;9:52–56.
77. Malyguine AM, Saadi S, Platt JL, Dawson JR. Human natural killer cells induce morphologic changes in porcine endothelial cell monolayers. Transplantation 1996;61:161–164.
78. Malyguine AM, Saadi S, Platt JL, Dawson JR. Differential expression of natural killer cell markers: human vs. baboon. Transplantation 1996;62:1319–1324.
79. Inverardi L, Samaja M, Motterlini R, Mangili F, Bender JR, Pardi R. Early recognition of a discordant xenogeneic organ by human circulating lymphocytes. J Immunol 1992;149:1416–1423.
80. Knechtle SJ, Halperin EC, Saad T, Bollinger RR. Prolonged heart xenograft survival using combined total lymphoid irradiation and cyclosporine. J Heart Transplant 1993;5:254–261.
81. van den Bogaerde J, Aspinall R, Wang M-W, Cary N, Lim S, Wright L, White D. Induction of long-term survival of hamster heart xenografts in rats. Transplantation 1991;52:15–20.
82. Alexandre GPJ, Gianello P, Latinne D et al. Plasmapheresis and splenectomy in experimental renal xenotransplantation. In: Hardy MA (ed) Xenograft 25. New York, NY: Elsevier Science Publishers, 1989;259–266.
83. Bellgrau D, Gold D, Selawry H, Moore J, Franzusoff A, Duke RC. A role for CD95 ligand in preventing graft rejection. Nature 1995;377:630–632.
84. Harland RC, Platt JL. Prospects for xenotransplantation of the liver. J Hepatol 1997;(In Press).
85. Chari RS, Collins BH, Magee JC, Kirk AD, Harland RC, McCann RL, Platt JL, Meyers WC. Treatment of hepatic failure with ex vivo pig liver perfusion followed by liver transplantation. N Engl J Med 1994;331:234–237.
86. Calne RY, White HJO, Herbertson BM, Millard PR, Davis DR, Salaman JR, Samuel JR. Pig-to-baboon liver xenografts. Lancet 1968;1:1176–1178.

87. Lawson JH, Platt JL. Molecular barriers to xenotransplantation. Transplantation 1996;62: 303–310.
88. Lawson JH, Daniels L, Platt JL. The evaluation of thrombomodulin activity in porcine to human xenotransplantation. Transpl Proc 1997;29:884–885 (Abstract).
89. Michaels MG, Simmons RL. Xenotransplant-associated zoonoses. Transplantation 1994;57: 1–7.
90. Patience C, Takeuchi Y, Weiss RA. Infection of human cells by an endogenous retrovirus of pigs. Nature Med 1997;3:282–286.

Tissue Engineering for Therapeutic Use 1.
Y. Ikada and Y. Yamaoka, editors.

Development of hybrid-type artificial immune organ by means of experimental evolutionary research method using bioceramics

Katsunari Nishihara
Department of Oral Surgery, Faculty of Medicine, University of Tokyo, Tokyo, Japan

The phylogenetic transition of branchiae into the lung as well as the migration of hemopoietic activity from the spleen to the bone marrow cavity demonstrate the concurrent evolutionary changes of organ structures and functions in vertebrates in response to environmental changes, such as the effect of gravity during terrestrialization from a water environment [1,2].

The present study introduces a hybrid-type artificial bone marrow chamber that uses sintered hydroxyapatite (HAP) and tricalcium phosphate (TCP) which is able to induce hemopoietic nest formation in conjunction with osteogenesis heterotopically in mammalian muscle, though not subcutaneous tissue [3–9]. This process is believed to proceed in biochambers by heterotopical hemopoiesis that are induced by the streaming potential of sintered HAP or TCP, which trigger genetic expression of mesenchymal cells to drive cytocaine of the bone morphogenetic protein (BMP). In order to investigate this hypothesis, the following preliminary experiments were carried out:

1. Transplantation of an HAP biochamber with BMP into the subcutis [10].
2. Transplantation of an HAP biochamber with a 10 μA current.
3. Measuring the streaming potential of sintered HAP using a physiological saline solution [10].
4. Transplantation of a titanium (Ti) mesh with a 10 μA current into the subcutis [10].

The preliminary results demonstrated that biomechanical stimuli were converted into streaming of the organic body fluid, ultimately resulting in the induction of a streaming potential [10]. Therefore, the author concluded that heterotopical hemopoiesis in the biochamber is induced by genetic expression of the mesenchymal cells. Based on this assumption, Ti artificial biochambers with a 10 μA current were developed. To further investigate this theory from a biomechanical perspective, a trilateral research method that integrates methodologies in morphology (Goethe), molecular biology (Delbrück) and physiology (Bernard) was developed. Based on this research method, the mechanisms associated with mor-

Address for correspondence: Dr Katsunari Nishihara, Department of Oral Surgery, Faculty of Medicine, University of Tokyo, 7-3-1 Hongo, Bunkyo-ku, Tokyo 113, Japan. Tel.: +81-3-3815-5411 (ext.) 3715. Fax: +81-3-5800-6832. E-mail: katsunari-n@pa.aix.or.jp phogenesis and function appear to originate from genetic expression, which are driven by biomechanical stimuli such as environmental factors.

An experimental evolutionary study, in which the trilateral research method is employed, was first proposed by the present author in 1995 [5,6]. In this study, biomechanical stimuli were applied to biomaterials in sintered hydroxyapatite chambers that have been implanted into various animal muscles during different phylogenetic stages. A hybrid-type artificial immune organ that induces hemopoietic bone marrow tissue was developed successfully not only in mammals, but also in chondrichthyes (sharks), which are relicts of the archetype that do not possess bone marrow tissue in their internal skeletons [11]. Using this method, an in vivo culture of an autogenous liver placed in a hydroxyapatite chamber which is connected to the femoral artery could also be established successfully in dogs [11].

The use and disuse theory proposed by Lamarck has been revived as a result of recent molecular genetic studies. Alberch reported that the recapitulation theory (Haeckel) could be explained by the heterochrony of genetic expression [12]. The use and disuse theory can also be explained by heterochrony, because all the functions of muscle cells and osteocytes are controlled through gene expression of mesenchymal and neural cells, which are triggered by physicochemical stimuli (also considered to be biomechanical stimuli) that affect the topical cells of the organism [10]. Based on this trilateral research, evolutionary changes in vertebrate morphology can be understood as a series of biomechanical events, as described by Lamarck [10]. Thus, evolutionary changes can be induced at the cellular level in the mesenchyma of heterospecies through biomechanical stimuli. This approach is referred to as an experimental evolutionary study [10].

The substance that defines the vertebrate structure is the hydroxyapatite-collagen complex skeleton that forms bone and teeth [5,6]. Vertebrates can be further distinguished by the presence of hemopoiesis in the bone marrow cavity of higher reptiles and mammals. This is not seen in lower vertebrates such as chondrocytes and cyclostomata. Structural differences also allow mammals to be distinguished from lower vertebrates. These consist of a highly evolved viscerocranium with gomphotic teeth and jawbones [5,6]. Combining the understanding of the importance of HAP in vertebrates with the effects of biomechanical stimuli, an experimental evolutionary research model can be developed at the cellular level [10]. Using such a model, a new technique in tissue engineering called the hybrid-type artificial immune organ has been developed [9]. If evolution is the product of biomechanical stimuli as proposed by Lamarck, heterospecific and heterotopical evolutionary changes should be possible to induce at the cellular level in the mesenchyma [9]. This paper investigates the correlation between the immunity system and evolutionary changes. Furthermore, this paper presents a new concept of immunology that involves a cytological digestion system [11], a new theory of biomechanics-responsive evolution [10] and a simple theory to explain biological reactions.

Materials and Methods

The following experimental evolutionary studies were carried out using artificial skeletal organs:

1) development of several kinds of artificial biochambers using:
   – artificial bone marrow chambers made of conventionally sintered HAP by Asahi Optical Co.,
   – artificial bone marrow chambers made of high-pressure sintering collagen-HAP composite [13] (by the National Institute for Research in Inorganic Material; collagen was extracted from cow skin),
   – artificial bone marrow chambers made of Ti mesh with a 10 μA current by the National Institute for Research in Inorganic Material [10], and
   – arterial biochamber for autogenous hepatic tissue culture,
2) implantation of the artificial bone marrow chambers into cyclostomata (hog-fish), chondrichthyes (dochishark), amphibian (xenops), aves (chicken) and mammals (dogs and Japanese monkeys),
3) based on the results of experiment 2, xenotransplantations between cyclostomata, chondrichthyes, amphibian, and mammals were carried out [14], and
4) transplantation of Ti biochamber with a 10 μA current into muscles of sharks and dogs.

Results

Artificial induction of hemopoiesis was carried out successfully in each of the newly developed chambers for all of the species. Figure 1 shows the leukocyte and lymphocyte induction around the Ti biochamber with a 10 μA current in a dog 4 months after implantation. Leukocyte production induced from undifferentiated mesenchymal cells by the electrical current around the Ti was observed.

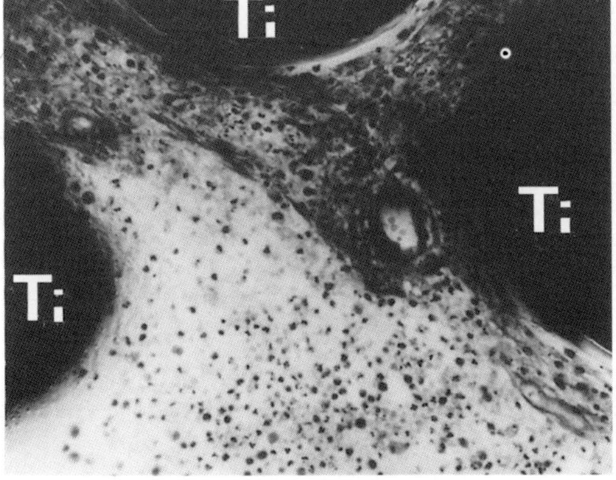

Fig. 1. Lymphocyte and leukocyte induction around Ti biochamber with 10 μA current. Ti: titanium.

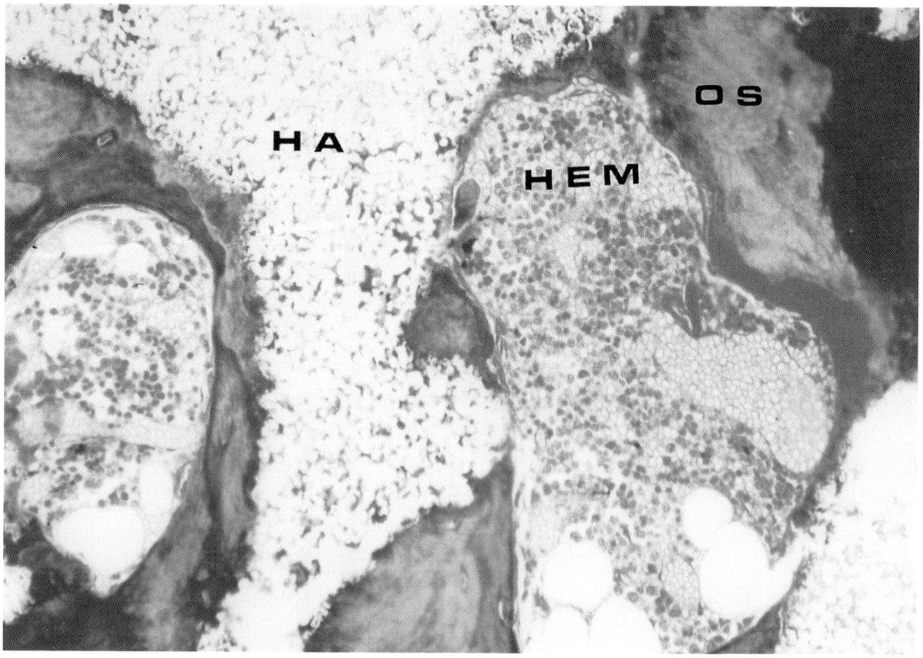

Fig. 2. Hemopoiesis with osteogenesis in HAP biochamber 12 months after implantation in the muscles of Japanese monkey. HA: hydroxyapatite; HEM: hematopoiesis; OS: osseous tissue.

Figure 2 shows induction 12 months after surgery of hemopoiesis in conjunction with osteogenesis by the HAP biochamber implanted in a Japanese monkey. Osteogenesis as well as hemopoiesis could be observed around the chamber. Figure 3 shows the autogenous liver tissue 2 months after surgery, cultured in the HAP biochamber connected to the femoral artery of a dog. However, in dog

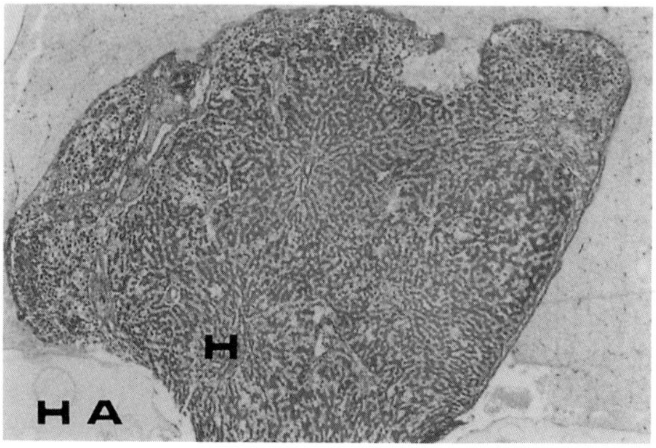

Fig. 3. Autogenous hepatic tissue cultured 2 months postoperation in HAP biochamber connected to dog femoral artery. HA: hydroxyapatite biochamber; H: hepatic tissue.

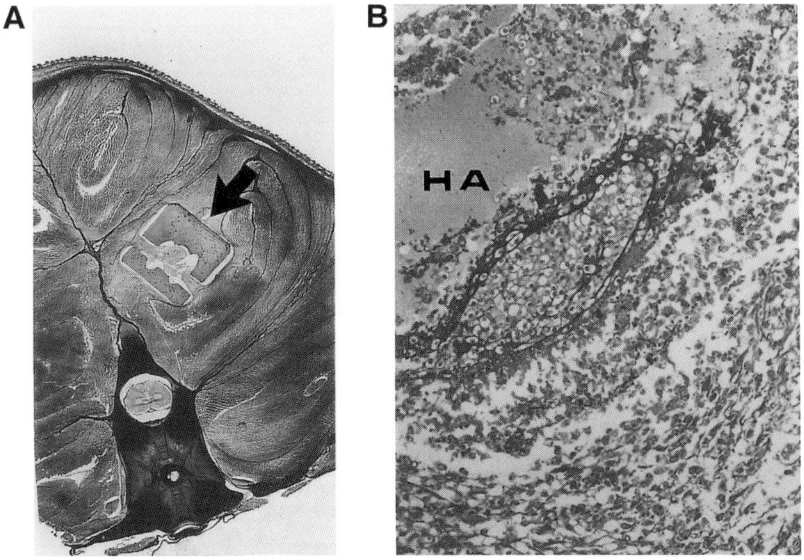

Fig. 4. **A:** Biochamber (arrow) of newly formed high pressure sintering composite collagen (cow skin)-HAP implanted in shark muscle. **B:** Osteoid formation with hemopoiesis is observed around biochamber.

muscles around the collagen-HAP biochambers, implanted marked tissue differentiation that resembled digestive tract formation could be observed. Figure 4A shows the collagen-hydroxyapatite artificial bone marrow chamber (arrow) implanted in shark muscle. Figure 4B shows hemopoiesis and osteoid formation 4 months after surgery around the hydroxyapatite implanted in the shark muscle. Figure 5A shows a cross-section of a shark (dochi). No bone marrow in the cartilaginous tissue around the spinal cord (arrow) is evident. Figure 5B shows a cross-section of a shark with a Ti biochamber with a 10 μA current. Figure 5C shows the histopathological findings associated with the dorsal cartilage with hemopoietic marrow induction (arrow) by the adjacent Ti biochamber. Reciprocal implantation of muscles between cyclostomata, chondrichthyes and xenops, mouse, and dogs was carried out successfully.

Discussion

Based on the phylogenetic transition of branchial smooth muscles into facial and mastication striated muscles, evolutionary changes in vertebrate organs are considered to reflect responses to environmental changes in biomechanical stimuli. Similarly, evolution of the branchiae into the lung and migration of hemopoietic activity from the spleen into the bone marrow cavity appear to coincide with changes in respiration from water to air and response to gravity associated with terrestrialization from a water environment (Fig. 6).

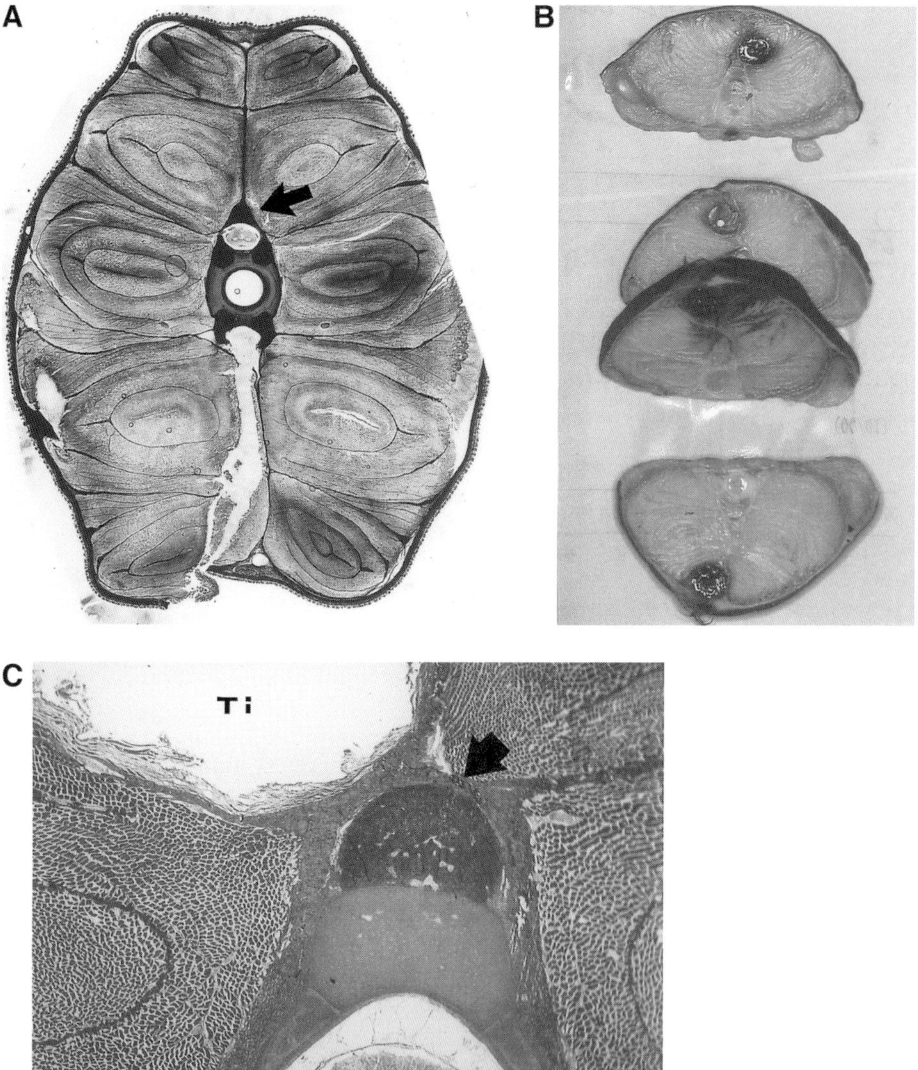

Fig. 5. **A:** Cross section of shark (dochi) for control, no hemopoietic marrow is seen in cartilage (arrow). **B:** Cross section of shark implanted with biochamber of Ti with current. **C:** Hematopoietic nest (arrow) induced by biochamber resembling bone marrow in shark vertebrate cartilage. Ti: biochamber of Ti with current.

The present series of studies were conducted based on the scientific categorization of vertebrates by Linne, the evolutionary theories of Lamarck, the morpho-

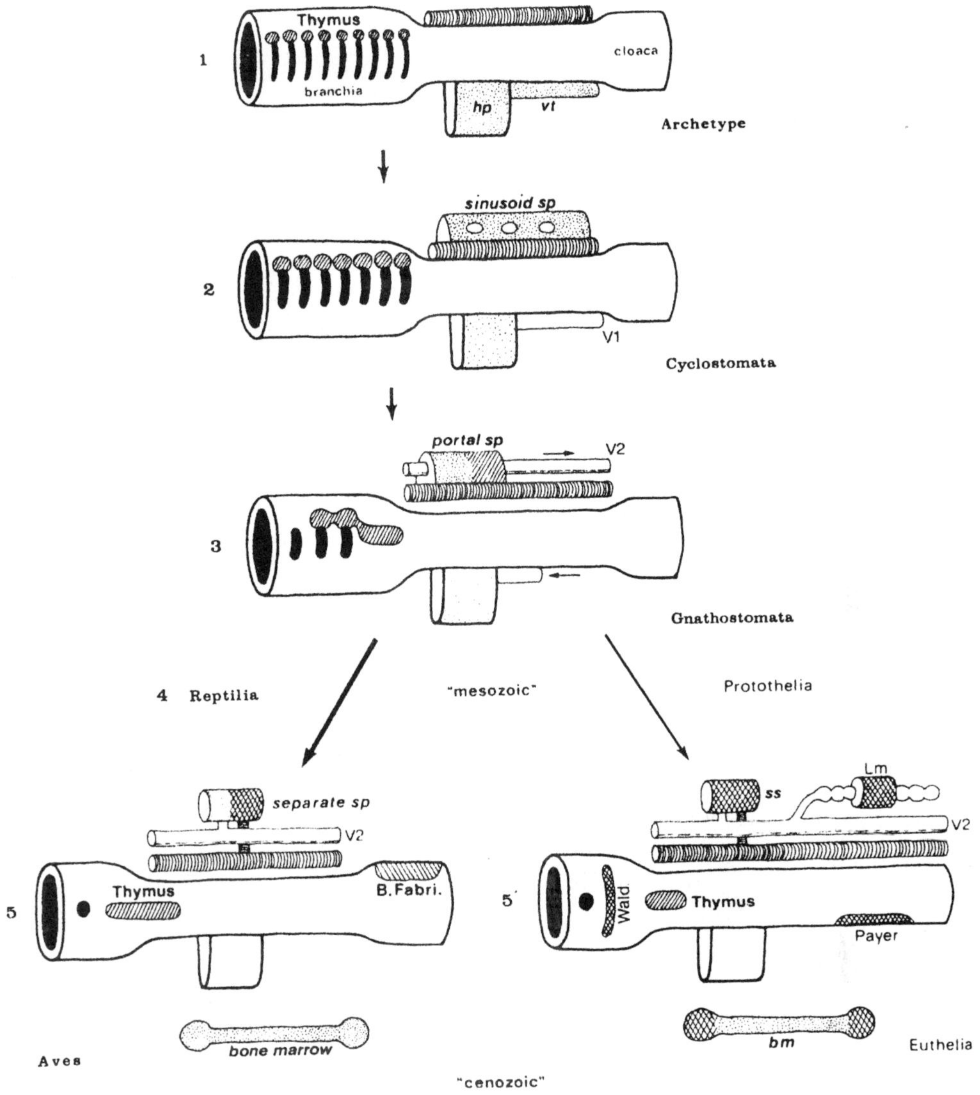

Fig. 6. Schematic diagram of evolution in vertebrate hemopoietic organs (from [17]).

logical studies by Goethe, the biogenetic law proposed by Haeckel, the biomechanical principles outlined by W. Roux [15], the morphology of organisms presented by S. Miki [16,17] and the molecular genetics developed by Delbrück [18]. The application of bioceramics to tissue engineering for therapeutic use was based on studies of phylogenetic transformation of osseous skeletal tissue in conjunction with hemopoiesis, i.e., the evolution of hematopoiesis in bone marrow. In the present study, morphology, physiology and biochemistry, and molecular genetics of remodeling were integrated with biomechanics to establish a trilateral research method for studying life sciences using biomaterials [5]. This trilateral research approach was used to perform an experimental evolutionary study of the mechanisms involved in evolutionary changes in vertebrates by applying bioceramics to archetype vertebrates that possess no inner HAP skeleton.

Conventional research has only studied the effects of biomaterials on the tissues surrounding the implanted artificial material. No marked histological changes appear to occur around the porous sintered HAP implanted into the subcutis. However, if the HAP is implanted into the muscle, marked heterotopical hemopoiesis in conjunction with osteogenesis occur around the HAP implant [19,20]. Thus, the response in subcutis appears to differ from that in muscle. Biomechanical stimuli tend to differ between the subcutis and muscle. Therefore, not only the material effects, but also the shape and functional effects, referred to hereafter as biomechanical stimuli, should be examined because the organism is a closed system in which internal pressure exists [21,22]. The stress distribution pattern around the implant is dependent on the shape of the implanted biomaterials. Cells with higher functions induced from undifferentiated mesenchyma are dependent on stress distribution patterns as well as the functional characteristics of the applied stress. In order to effectively use biomaterials in clinical applications, the material, shape and functional effects must be clearly understood [21,22]. Through a trilateral approach using biomaterials, tissue engineering of hybrid-type artificial immune organs can be developed. Furthermore, by introducing gravity as well as low-current energy, a simple theory to explain biology can also be established. New immunological concepts that consider cytological respiration, metabolism and digestion in conjunction with cell differentiation can also be introduced [11].

Based on the successful induction of hemopoiesis and osteoid formation by a collagen-HAP biochamber of cow skin, chondrichthyes and cyclostomata do not appear to demonstrate any major histocompatibility antigen. Autogenous hepatic tissue culture was carried out successfully using a HAP biochamber that was connected to the femoral artery of a dog (Fig. 3). All epithelial organs derived from the endoderm and mesoderm showed epithelial-mesenchymal interaction. Due to HAP being a product of the mesenchyma, the use of a HAP chamber for in vivo culture of an autogenous endodermal epithelial organ appears to be the most appropriate biochamber [11].

In 1965 T. McCulloch developed the colony forming a unit-spleen method during lethal-dose irradiation exposure experiments using mice, one of which he did not irradiate on the tail. That mouse did not die, despite exposure to a lethal dose of radiation over the whole body except the tail [23]. The mouse appeared to recover due to proliferation of stem cells that migrated from the tail bone marrow. Thomas successfully applied the same method to human bone marrow transplantation in order to treat hematopoietic disorders [24], for which he was awarded the 1990 Nobel prize in medicine and physiology.

Based on the colony forming unit-spleen method as well as research on bone marrow cells [25—28] and phylogenetic research on evolution of the hemopoietic organ [16,17,29] (Fig. 6), the concept of a hybrid-type artificial bone marrow chamber was developed [30,31]. Because all mesenchymal cells that comprise the body of vertebrates have common genetic codes, undifferentiated mesenchymal cells can differentiate into various types of cells that comprise the skeleton, i.e., osseous and cartilaginous tissue, muscle, blood, reticuloendothelial, lipid, tendon, endothelial cells, and fascia.

The present results suggest that the trigger for hemopoiesis induction in conjunction with osteogenesis is a streaming potential [31—34] which is converted from fluid flow that occurs due to repeated optimal loading. Cellular functions can be categorized as follows: 1) metabolism, 2) skeletal construction, 3) muscle activity, 4) planocytes and phagocytes for cytological respiration and digestion, 5) signal transmission of neural cells, 6) absorption of nutrients and oxygen, 7) excretion of catabolites, and 8) support of metabolites for epithelial cells. Almost all cells are capable of cell division and remodeling. Osseous cells function as a support against mechanical stress. They form the basic skeleton of the organism in response to biomechanical stimuli. Almost all cellular functions are ultimately carried out through genetic expression of each cell.

Morphology, metabolism and remodeling are all products of cellular function. These cellular functions are also products of genetic expression of cells. Through trilateral research that integrates morphology, metabolism and molecular genetics of remodeling with responses to biomechanical stimuli, these three categories can be studied as the same molecular genetic phenomenon with three different features.

Thus, genetic expression appears to be under the control of the general biomechanical stimuli. The present results demonstrate that the migration of hemopoiesis from the spleen into the bone marrow cavity occurred as a result of gravity that became a factor during the terrestrialization period of evolution. The only difference between archetype vertebrates and higher vertebrates is blood pressure. Mammals have substantially higher blood pressure than sharks. By loading a mechanical stress of 1 g, blood pressure should dramatically increase in sharks that have very low blood pressure in water, with approximately 1/6 g.

During terrestrialization, archetype vertebrates (sharks) initially experienced difficulty in breathing due to the lack of respiration without water and writhed in response to the suffocation. As a result, blood pressure increased. During surgery, bleeding was not apparent in most areas of shark muscle. Thus, gravity appears to be the primary influence on the elevation of blood pressure during terrestrialization. This in turn induced a higher streaming potential [31] than that in the water environment. Therefore, hemopoietic nest migration appears to be a direct consequence of the introduction of gravity. Furthermore, these studies demonstrate that evolutionary changes in morphology result from biomechanical-physicochemical stimuli such as gravity and calcium ions. Figure 7A shows a schematic diagram of the neoteny of the ascidia. Figure 7B shows the artificially induced neoteny of the maboya (ascidia) in artificial sea water with a low density of calcium ions or with gadolinium ions instead of calcium [35]. The role of grav-

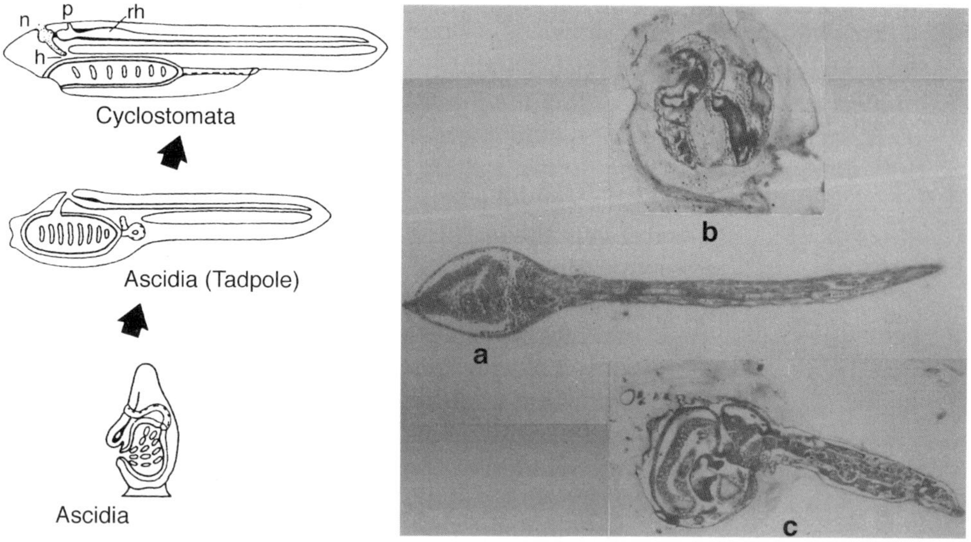

Fig. 7 **A:** Schematic diagram of neoteny of ascidia to vertebrate. **B:** Artificial induction of neoteny of maboya (ascidia). a. Larva (tadpole type); b. imago; c. imago with tail.

ity in the induction of major histocompatibility antigens is further demonstrated through experimental transplantation between sharks, hogfish and xenops, mice, rats and dogs [14]. Therefore, not only hemopoietic nest migration, but also major histocompatibility antigens appear to be induced by gravity through the genetic expression of mesenchymal cells [14].

Summary

The present experimental evolutionary study investigated the artificial induction of bone marrow hemopoiesis in heterospecies using chondrichthyes and cyclostomata, which have no osseous bone in their inner skeleton, and therefore represent the phylogenetical stage of archetype vertebrates. The major histocompatibility antigens appear to have differentiated concomitant with migration of hemopoiesis into the bone marrow during terrestrialization. Based on these results, this study presents a new concept of the immune system, a new theory of biomechanics-responsive evolution, as well as a simple theory to explain biological reactions.

Acknowledgements

The present study was supported in part by a Grant-in-Aid for Scientific Research on Priority Area (1) (No. 08233102) and a Grant-in-Aid for Co-operative Research (A) (No. 07309007) from the Ministry of Education, Science and Culture, Japan.

References

1. Miki S. Über die genetische Beziehung der Milz zu den Gefässen des Magens besonders zu seinen sekundären Venen beim Riesensalamander. Acta Anat Nipp 1963;38:140–155.
2. Hartman A. Die Entwicklung der Milz vom ersten Autreten der Anlage bis zu Differenzierung zum fertigen Organ. I Amphibien (Pleurodeles) Z mikr-anat Forsch 1933;34:553–655.
3. Nishihara K. Evolution of mechanical supportive organ system and medical application of biomaterials. J Jpn Soc Biomater 1993;11(5):31–40.
4. Nishihara K. Phylogeny and biomechanics – biomechanics concerning vertebrate evolution. J S Mech Eng 1994;(940–5):84–85.
5. Nishihara K. What is the viscerocranium from the standpoint of vertebrate evolution? J Oromax Biomech 1995;1(1):73–78.
6. Nishihara K. The basic construction of vertebrates, structural defects in the human body and a new concept of the immune system. J Oromax Biomech 1995;1(1):79–86.
7. Nishihara K, Tange T, Tokumaru H, Yanai A, Hirayama Y. Study on developing artificial bone marrow made of sintered hydroxyapatite chamber. Bioceramics 1993;5:131–138.
8. NishiharaK, Tange T, Hirota K, Kawase K. Development of hybrid type artificial bone marrow using sintered hydroxyapatite. Biomed Mat Eng 1994;4(1):61–65.
9. Nishihara K, Tanaka J. Succssful inducement of hybrid type artificial bone marrow using bioceramics in various vertebrates. Bioceramics 1996;9:69–72.
10. Nishihara K, Tanaka J, Hirota K. Study on mechanism of biomechanics-inducing evolution in vertebrates by means of experimental evolutionary research method. Jpn J Oral Diag/Oral Med 1996;9(2):232–249.
11. Nishihara K, Tange T, Matsuda R, Tanaka J, Hirota K, Kabasawa H. Successful inducement of artificial bone marrow and the new concept of immune system. Jpn J Oral Diag/Oral Med 1996;9(2) 217–231.
12. Alberch P. Heterochrony; Pattern or process? Biodiversity and Evolution, the 10th International Symposium on Biology in Conjunction with the Awarding of the International Prize of Biology 1994;26–27.
13. Hirota K, Nishihara K, Tanaka H. Pressure sintering of apatite-collagen composite. Biomed Mat Eng 1993;3(3):147–151.
14. Nishihara K, Kabasawa H. Relation between gravity and cell differentiation in vertebrates – a new concept of immunology. J Oromax Biomech 1996;2(1):19–22.
15. Roux W. Gesammelte Abhandlungen der Entwickungsmechanik der Organismus. Leipzig: Wilhelm Engelmann, 1895.
16. Miki S. Natural History of Morphology, 1st edn. Ubusuna Publishing Co., Tokyo, 1991.
17. Miki S. General Morphology of the Organisms – Basic Forms and Metamorphosis, 1st edn. Tokyo: Ubusuna Publishing Co., 1993.
18. Delbrück M. A physicist looks at biology. Trans Connecticut Acad Arts Sci 1949;38:173–190.
19. Nishihara K. Successful Inducement of Hemopoietic Nest in Artificial Bone Marrow Chamber. Abstracts of the 10th World Congress of the International Society for Artificial Organs in Conjunction with the International Faculty for Artificial Organs 1995;143.
20. NishiharaK, Tange T, Tokumaru H, Yanai A, Hirayama Y. Study on developing artificial bone marrow made of sintered hydroxyapatite chamber. Bioceramics 1992;5:131–138.
21. Nishihara K, Akagawa T. Comparative studies on periodontal tissue around new type artificial roots made of zirconium oxide, titanium, and hydroxyapatite. Phosphor Res Bull 1991;1: 179–184.
22. NishiharaK, Jiang L, Kobayashi T, Yanai A, Nakagiri S. Studies on functional effect of hydroxyapatite artificial root upon surounding tissues – new concept for bone-bioceramics jointing system. Bioceramics 1992;5:333–342.
23. Till J, McCulloch EA. A direct measurement of the radiation sensitivity of normal mouse bone marrow cells. Radiat Res 1961;14:313–322.

24. Thomas ED, Storb R. Technique for human marrow gafting. Blood 1970;36(4)(October): 507–515.
25. Mori M, Miwa A, Mori S. Regeneration of bone marrow in grafted decalcified femur of C3H mice; the role of bone and bone matrix. Recent Adv RES Res 1985;25:36.
26. Friedenstein A J. Precursor cells of mechanocytes. Rev Cytol 1976;47:327–355.
27. Ohgushi H. Heterotopic osteogenesis in porous ceramics induced by marrow cells. J Orthop Res 1989;7:568–578.
28. Ernst PB, Befus AD, Bienenstock J. Leukocytes in the intestinal epithelium: an unusual immunological compartment. Immunol Today 1985;6(2):50–55.
29. Nelsen OE. Comparative Embryology of the Vertebrates, 1st edn. New York: The Blakiston Company, Inc., 1953;733–736.
30. Nishihara K. Approach for hybrid-type artificial bone marrow inducemet. Jpn J Art Org 1995;24(1):6–12.
31. Nishihara K, Tange T, Matsuda R, Seno H, Yanai A, Fujigi K, Tanaka J, Hirota K. Development of hybrid-type artificial organs by means of experimental evolutionary method and concept of the immune system. Jpn J Art Org 1996;25(3):753–758.
32. Pollacck S R, Petrov N, Salzstein R, Barnkov G, Blaboeva R. An anatomical model for streaming potentials in osteons. J Biomechanics 1984:17(8):628–637.
33. Fukuda E, Yasuda I. On the piezoelectric effect of bone. J Phys Soc Jpn 1957;12:1158–1162.
34. Basset C. Biophysical principles affecting bone structure. In: Bourne G (ed) The Biochemistry and Physiology of Bone, vol 3, 2nd edn. New York: Academic Press, 1971;1–76.
35. Nishihara K, Sato Y, Morisawa M. Morphology of the viscerocranium and evolution of vertebrates – evidence of experimental neoteny using ascidia. J Oromax Biomech 1996;2(1):16–18.

Tissue Engineering for Therapeutic Use 1.
Y. Ikada and Y. Yamaoka, editors.

Growth factors as local regulatory factors in fracture repair

Seiya Jingushi and Yukihide Iwamoto
Department of Orthopaedic Surgery, Kyushu University, Fukuoka, Japan

Keywords: bone, cartilage, FGFs, fracture, growth factors, TGFβ, tissue engineering.

Introduction

Fracture repair begins with bleeding at the fracture site, followed by inflammatory reaction. The process continues to formation of reparative granuloma, called a callus, in which osteogenesis and chondrogenesis occur simultaneously and/or sequentially. Histological examinations of the fracture repair have shown that the repair process is well-organized [1]. However, it is not clear how the fracture repair is regulated or what the regulators for the cellular events are.

Recent studies suggest that wound healing is regulated by growth factors, proteins that affect cell proliferation, differentiation, and protein synthesis by autocrine or paracrine mechanism. In vitro studies have shown that growth factors including transforming growth factor beta 1 (TGFβ1) and basic fibroblast growth factors (bFGF) affect several cell types found in the fracture callus, including chondrocytes and osteoblasts [2,3]. Additionally, large amounts and many kinds of growth factors can be extracted from bone and cartilage tissues [4–8]. This suggests that the growth factors regulate fracture repair.

In this paper, cellular events which occur during fracture repair are described first. Expression of growth factors in the fracture callus is then shown. Finally, whether addition of exogenous growth factors can modulate the repair process or not is shown in order to clarify the role of the growth factors.

Cellular events during fracture repair

Many cellular events occur during fracture repair including immediate injury response, intramembranous ossification, cartilage formation, endochondral ossification and bone remodeling. These cellular events appear in the callus simultaneously and/or sequentially (Fig. 1).

Bone forms in the callus by two distinct processes: intramembranous ossification and endochondral ossification. Bone formed by intramembranous ossification does not restore bone continuity, but forms an external "buttress" for bone

---

Address for correspondence: Dr Seiya Jingushi MD PhD, Department of Orthopaedic Surgery, Kyushu University 3-1-1 Maidashi, Higashi-ku, Fukuoka 812, Japan. Tel.: +81-92-642-5489. Fax: +81-92-642-5507. E-mail: jingushi@ortho.med.kyushu-u.ac.jp

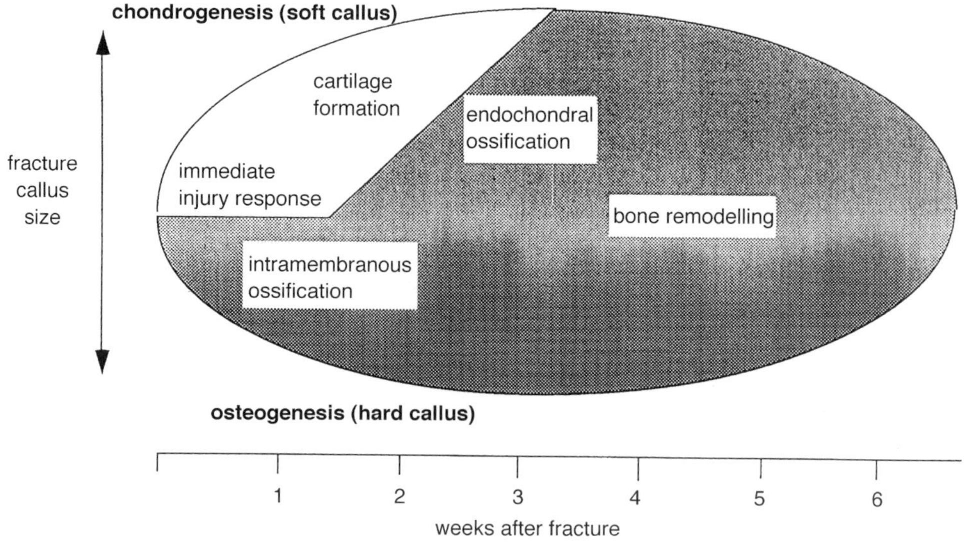

Fig. 1. Cellular events during fracture repair using rat fracture model. Many cellular events occur during fracture repair simultaneously or sequentially. This schema indicates the relationship between each cellular event. This rat fracture model takes approximately 6 weeks to heal. The height of the schema represents the size of callus. The callus increases until approximately 2–3 weeks after the fracture by tissue generation, and then decreases through bone remodeling mechanism. The callus can be divided into soft callus (fibrous and cartilage tissue) and hard callus (bone tissue). Just after fracture, immediate injury response occur at the fracture site, and intramembranous ossification (periosteal reaction) initiates adjacent to the fracture site. Also, in the soft callus, cartilage tissue appear. In the hard callus, intramembranous ossification proceeds, and newly formed trabecular bone is generated and grows. Cartilage tissue is replaced by bone tissue through endochondral ossification.

that forms subsequently. Cartilage, which forms at the fracture site, is replaced by endochondral ossification which bridges the fracture site, restoring bone continuity. Normal bone anatomy is restored by the remodeling of the new bone.

Histology

Detailed histological changes using rat femoral fracture model [9] are described as follows [1] (Fig. 2):

Immediate injury response
Immediately after fracture, a hematoma forms at the fracture site and extends along the cortex above the periosteum. Soft tissue surrounding the fracture site is invaded by inflammatory cells, as the fibrous clot is organized into granulation tissue.

Intramembranous ossification
Shortly after the fracture, subperiosteal cells proliferate and the periosteum thickens near the fracture site. Newly formed osteoid is also seen at the same time on

the cortex. As intramembranous ossification proceeds, mature trabecular bone begins to form and is visible by day 7.

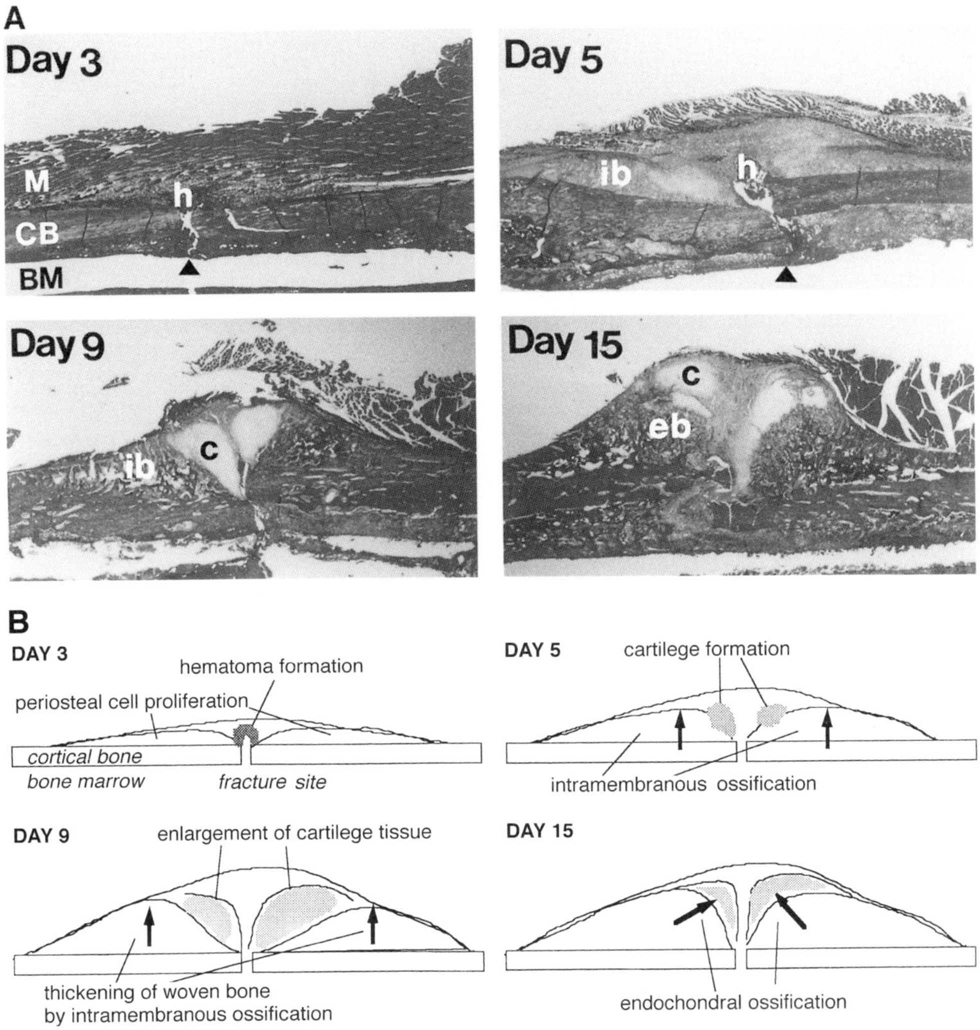

Fig. 2. Histological changes of rat femoral fracture calluses. Photographs (**A**) and each schema (**B**). Each photograph shows half of the callus on days 3, 5, 9 and 15 after the fracture. They are stained by Masson Trichrome. On day 3, hematoma is observed at the fracture site, and there are many mesenchymal cells around the hematoma. Periosteal reaction appears adjacent to the fracture site long along the cortex. On day 5, woven bone is generated through intramembranous ossification. Cartilage tissue appear adjacent to the newly formed bone. On day 9, most cells in the soft callus are chondrocytes. Thick trabecular bone is generated like an external "buttress". On day 15, endochondral ossification is observed in the junction between cartilage tissue and trabecular bone. The endochondral ossification proceeds until bone bridges the fracture site. M: muscle; h: hematoma; BM: bone marrow; ib: intramembranous ossification; c: cartilage; eb: endochondral ossification.

Cartilage formation

Cartilage formation occurs adjacent to the newly formed trabecular bone by intramembranous ossification. It is first seen on day 5. Depending on the progression of the repair process, the type of chondrocytes observed in the cartilage tissue varies from small immature chondrocytes to large hypertrophic chondrocytes surrounded by rich matrix indicating the chondrocyte differentiation.

Endochondral ossification

It is first seen on day 9 in hypertrophic cartilage adjacent to the newly formed trabecular bone. By day 15, it occurs along the entire junction between cartilage tissue and newly formed trabecular bone, or between soft callus (fibrous and cartilage tissue in the callus) and hard callus (bone tissue in the callus). Endochondral ossification proceeds until bone bridges the fracture site. The process is seen between 28 and 35 days after fracture in this model.

Cell proliferation during fracture repair

Localization of cell proliferation in the fracture calluses was investigated using immunohistochemistry and monoclonal antibodies against proliferating cell nuclear antigen (PCNA) [10]. PCNA is a coenzyme for polymerase d and is a specific protein for proliferating cells.

A total of 36 h after fracture, many proliferating cells were observed in the whole callus. It included the periosteal cells where intramembranous ossification occurred and the mesenchymal cells at the fracture site. On day 3, there were many proliferating periosteal cells and premature osteoblasts in the newly formed trabecular bone. On day 7, many premature chondrocytes were proliferating. Endochondral ossification appearers on day 10 or later. In the endochondral ossification front, premature osteoblasts proliferate, although hypertrophic chondrocytes do not.

Cell differentiation during fracture repair [1]

The fracture callus can be divided into regions of bone formation (hard callus) and cartilage formation (soft callus). Gene expression for matrix proteins is examined separately in each region in order to investigate cell differentiation in the fracture callus. Expression of extracellular matrix protein genes vary with the progression of repair and differ between hard and soft calluses. Messenger ribonucleic acids (mRNAs) for osteonectin, alkaline phosphatase, and type I procollagen are detected in the hard callus at maximal levels during endochondral ossification and bone remodeling (day 15) and at 50% maximal levels during intramembranous bone formation (day 7). mRNAs for type II procollagen and proteoglycan core protein are detected at maximal levels in the soft callus during chondrogenesis (day 9). Osteocalcin gene expression is detected in the hard callus during endochondral ossification and remodeling, but not during intramembranous bone formation or at any time in the soft callus. Expression of cartilage and bone-related genes correlated with the temporal sequence of histological changes, suggesting transcriptional regulation of gene expression during fracture repair.

Growth factors in fracture repair

Localization of growth factors in fracture calluses

TGFβ1 [11], bFGF [12], PDGF [13], BMP-4 [14], and IGF-I [15] are expressed in the callus. Localization of the growth factors are described in each cellular event (Fig. 3).

Immediate injury response
Extracellular TGFβ1 was localized in the hematoma early after fracture and persisted in this region up to 10 days. It extended along the periosteum. TGFβ1 was released by platelets into the fracture hematoma [16]. Subperiosteal injections of TGFβ1 on uninjured bone caused bone and cartilage formation like a callus [17,18]. TGFβ1 from platelets released early after fracture seemed to play a role in triggering the repair process. PDGF was also released from platelets and is thought to function as a trigger. The mesenchymal cells at the fracture site expressed bFGF.

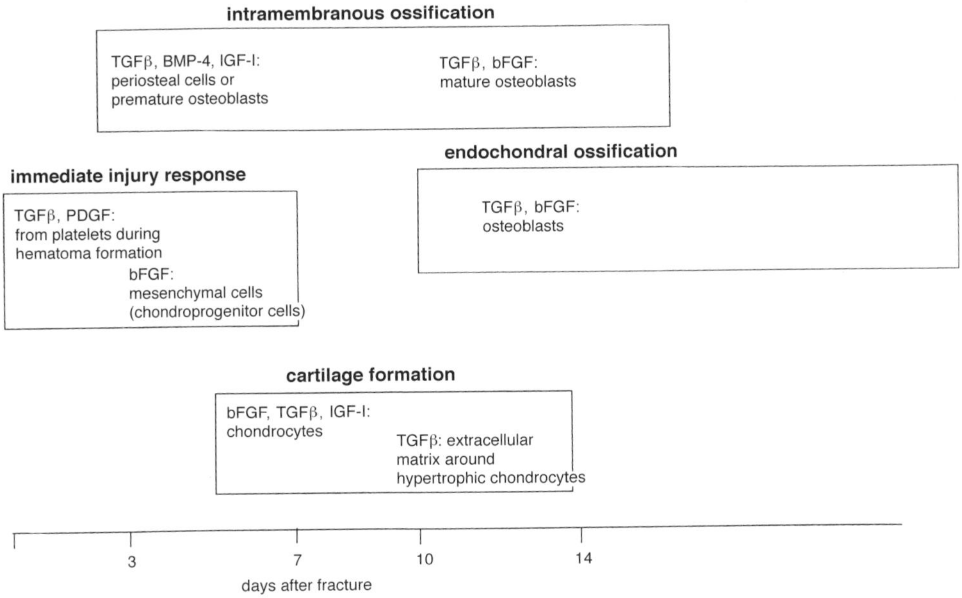

Fig. 3. Expression of growth factors in each cellular events.

Intramembranous ossification
About 2 or 3 days after fracture, intracellular TGFβ1 is observed in immature osteoblasts in the newly formed trabecular bone. BMP-4 and IGF-I are also expressed in the immature osteoblasts early after fracture. TGFβ1, BMP-4 and IGF-I are thought to regulate proliferation and early differentiation of the osteoblast in that region. From 10 to 14 days after fracture, TGFβ1 and bFGF, though neither BMP-4 nor IGF-I, are expressed in mature osteoblasts. They seem to regulate terminal differentiation and matrix protein synthesis of osteoblasts.

Cartilage formation
Chondroprogenitor cells are included in mesenchymal cells which appear at the fracture site shortly after injury. They express bFGF as described above. Proliferating chondrocytes express TGFβ1, bFGF and IGF-I, though not hypertrophic chondrocytes. TGFβ1 stimulates the chondrocyte phenotypes of periosteal cells [19]. Exogenous TGFβ1 and bFGF on uninjured bones causes cartilage formation as well as bone formation [17,18]. TGFβ1 and bFGF which are expressed in chondroprogenitor cells or chondrocytes are thought to regulate chondrocyte proliferation and differentiation during fracture repair.

Endochondral ossification
TGFβ1 is observed in the extracellular matrix around hypertrophic chondrocytes. The TGFβ1 may be an angiogenic factor to induce capillary vein invasion and a chemotactic factor for chondroclasts. TGFβ1 and bFGF are expressed by the osteoblasts in the endochondral ossification front. These growth factors seem to regulate the osteoblast differentiation or matrix protein synthesis during endochondral ossification.

Gene expression for TGFβ1 [11]

In order to localize TGFβ1 gene expression, the fracture callus is microdissected into soft (fibrous and cartilaginous) and hard (osseous) callus and RNA is separately extracted from each. TGFβ1 gene expression is detected both in the soft and hard calluses. It is interesting that the gene expression varies depending on the change of matrix protein gene expression both in the soft and hard calluses. In the soft callus, message levels for TGFβ1 peaks at 13 days after fracture. This broad peak corresponds with chondrogenesis or chondrocyte differentiation. In the hard callus, the message levels peak on days 5 and 15. The two peaks correspond with intramembranous and endochondral ossification, respectively. This suggests that TGFβ1 is one of the important regulators for chondrogenesis.

Modulation of fracture repair by exogenous growth factors (Table 1)

Growth factors which are expressed in the fracture callus are thought to regulate cellular events which occur during fracture repair. The next question is how

Table 1. Effects of exogenous FGFs and/or TGFβ.

A. On fracture repair

Injected growth factors	Single/ repeated	Fractured bone	Modulation of fracture calluses	Reference
FGF	repeated	rat femur	enlargement of cartilage tissue	20
bFGF	repeated	rat femur	enlargement of cartilage tissue	20
bFGF and TGFβ	repeated	rat femur	enlargement of soft callus, fibrous tissue and cartilage tisssue	29
TFGβ	repeated	rat and rabbit femur	improvement of mechanical properties of fracture callus	25, 26
bFGF	single	rat fibula	improvement of mechanical properties of fracture callus	24

B. On uninjured bone

Injected growth factors	Single/ repeated	Injection site	Generated tissues	Reference
aFGF	repeated	rat femur	cartilage and bone tissue	21
bFGF	repeated	rat femur	cartilage and bone tissue	21
TGFβ	repeated	rat femur	cartilage and bone tissue	17, 18
TGFβ	repeated	rat frontal prarietal bone	cartilage and bone tissue	28
TGFβ	repeated	rat calvariae	bone tissue	27

important these growth factors are as local regulators for the repair process. One of the ways to investigate this question is to evaluate whether exogenous growth factors can modulate the repair process. This can also be used to investigate whether these growth factors are candidates for the clinical use for treating pathological healing or, perhaps, to fasten or strengthen the bone healing.

FGFs on fracture repair

Repeated injections of aFGF into a fracture site in rat closed femoral fractures causes a marked increase in the size of the cartilaginous soft callus [20]. Gene expression for matrix proteins, relatively specific for cartilage tissue, are inhibited by the aFGF injections. Exogenous aFGF causes the stimulation of chondroprogenitor cell proliferation, and suppresses differentiation. The aFGF injections do not influence bone formation or mechanical properties of the callus. In the same models, repeated bFGF injections also cause enlargement of cartilage tissue, and do not influence bone formation significantly. This may be due to the method of FGF administration onto the callus. However, both aFGF and bFGF can stimulate intramembranous bone formation on uninjured rat femoral bone [21]. Several reports show that bFGF stimulates endosteal bone formation [22] and enhances bone-graft incorporation [23]. Additionally, a single administration of bFGF on a rat fibular open fracture stimulates fracture repair in streptozotocin-diabetic rats in which the repair process is prolonged compared with normal fracture repair [24].

TGFβ on fracture repair

Local injections of TGFβ are reported to increase rat or rabbit fracture repair [25,26]. The injection results in an increase in mechanical properties. Subpriosteal injections of TGFβ on uninjured femoral bone cause bone and cartilage formation [17,18]. TGFβ is also reported to induce intramembranous ossification in neonatal rat calvariae and endochondral ossification in adult rats [27,28].

Combination of growth factors

Repeated injections of TGFβ and bFGF result in significant modulation of the fracture repair [29]. This causes a significantly enlarged soft callus. The composition of the callus depends on the dose ratio of TGFβ and bFGF. More bFGF causes more cartilage tissue. More TGFβ causes more fibrous tissue. Exogenous TGFβ and bFGF could control the tissue type in the fracture callus.

Exogenous FGFs and/or TGFβ modulate fracture repair process indicating the cells which react with the growth factors appear during the process. The growth factors modulate chondrogenesis and osteogenesis during the repair process indicating that the growth factors regulate such cellular events. The growth factors may regulate the repair process co-operatively rather than independently.

Summary

Many growth factors are expressed in the fracture callus in various regions and at different times after the fracture. These regulate cell proliferation and differentiation in chondrogenesis and osteogenesis during fracture repair as local factors. Exogenous growth factors can modulate chondrogenesis and osteogenesis. The effects depend on the type of growth factors and on the injection way. Growth factors which are expressed during fracture repair can be candidates for improving pathological fracture repair. With respect to tissue generation, these growth factors can also induce generation of new cartilage and bone tissue. These potent peptides are thought to be beneficial in resolving many orthopaedic problems including pathological fracture repair, bone and cartilage defects.

References

1. Jingushi S, Joyce ME, Bolander ME. Genetic expression of extracellular matrix proteins correlates with histological changes during fracture repair. J Bone Min Res 1992;7(9):1045–1055.
2. Baird A, Bohlen P. Fibroblast growth factors. In: Sporn MB, Roberts AB (eds) Peptide Growth Factors and their Receptors. New York: Spring-Verlag, 1991;392–402.

3. Roberts AB, Sporn MB. Transforming growth factor-βs. In: Sporn MB, Roberts AB (eds) Peptide growth factors and their receptors. New York: Spring-Verlag, 1991;435–448.
4. Hauschka PV, Mavrakos AE, Iafrati MD, Doleman SE, Klagsbrun M. Growth factors in bone matrix: isolation of multiple types by affinity chromatography on heparin-Sepharose. J Biol Chem 1986;261:12665–12674.
5. Hamerman D, Sasse J, Klagsbrun M. A cartilage-derived growth factor enhances hyaluronate synthesis and diminishes sulfated glycosaminoglycan synthesis in chondrocytes. J Cell Physiol 1985;127:317–322.
6. Hill DJ, Logan A, Ong M, Sousa DD, Gonzalez AM. Basic fibroblast growth factor is synthesized and released by isolated ovine fetal growth plate chondrocytes: potential role as an autocrine mitogen. Growth Factors 1992;6:277–294.
7. Ellingsworth LR, Breenan JE, Fok K, Rosen DM, Bentz H, Piez KA, Seyedin SM. Antibodies to the N-terminal portion of cartilage-inducing factor A and transforming growth factor β. J Biol Chem 1986;261:12362–12367.
8. Thompson NL, Flanders KC, Smith JM, Ellingsworth LR, Roberts AB, Sporn MB. Expression of transforming growth factor-β1 in specific cells and tissues of adult and neonatal mice. J Cell Biol 1989;108:661–669.
9. Bonnarens F, Einhorn TA. Production of a standard closed fracture in laboratory animal bone. J Orthop Res 1984;2:97–101.
10. Iwaki A, Jingushi S, Oda Y, Izumi T, Shida J, Tsuneyoshi M, Sugioka Y. Localization and quantification of proliferating cells during rat fracture repair: detection of proliferating cell nuclear antigen by immunohistochemistry. J Bone Min Res 1997;12(1):96–102.
11. Joyce ME, Jingushi S, Bolander ME. Transforming growth factor β in the regulation of fracture repair. Orthop Clin N Am 1990;21(1):199–209.
12. Jingushi S, Joyce ME, Bolander ME. Basic FGF in rat fracture repair. Orthop Trans 1993;17(3): 713–714.
13. Bolander ME. Regulation of fracture repair by growth factors. Proc Soc Exp Biol Med 1992; 200:165–170.
14. Nakase T, Nomura S, Yoshikawa H, Hshimoto J, Hirota S, Kitamura Y, Oikawa S, Ono K, Takaoka K. Transient and localized expression of bone morphogenetic protein 4 messenger RNA during fracture healing. J Bone Min Res 1994;9(5):651–659.
15. Edwall D, Prisell PT, Levinovitz A, Jennische E, Norstedt G. Expression of insulin-like growth factor I mRNA in regenerating bone after fracture: influence of indomethacin. J Bone Min Res 1992;(7):207–213.
16. Assoian RK, Sporn MB. Type-β transforming growth factor in human platelets: release during platelet degranulation and action on vascular smooth muscle cells. J Cell Biol 1986;102: 1217–1223.
17. Joyce ME, Roberts AB, Sporn MB, Bolander ME. Transforming growth factor β and the initiation of chondrogenesis and osteogenesis in the rat femur. J Cell Biol 1990;110:2195–2207.
18. Joyce ME, Jingushi S, Roberts AB, Sporn MB, Bolander ME. Transforming growth factor-β initiates cartilage and bone formation in vivo. In: Cohn DV, Glorieux, Martin TJ (eds) Calcium Regulation and Bone Metabolism. Amsterdam: Elsevier Science Publishers,1990;317–323.
19. Izumi T, Scully S, Heydemann A, Bolander ME. Transforming growth factor β1 stimulates type II collagen expression in cultured periosteum-derived cells. J Bone Min Res 1992;7(1): 115–121.
20. Jingushi S, Heydemann A, Kana SK, Macey LR, Bolander ME. Acidic fibroblast growth factor injection stimulates cartilage enlargement and inhibits cartilage gene expression in rat fracture healing. J Orthop Res 1990;8:364–371.
21. Jingushi S. Unpublished data.
22. Mayahara H, Ito T, Nagai H, Miyajima H, Tsukuda R, Taketomi S, Mizoguchi J, Kato K. In vivo stimulation of endosteal bone formation by basic fibroblastic growth factor in rats. Growth Factors 1993;9:73–80.

23. Wang JS, Aspenberg P. Basic fibroblast growth factor enhances bone-graft incorporation: dose and time dependence in rats. J Orthop Res 1996;14:316–323.
24. Kawaguchi H, Kurokawa T, Hanada K, Hiyama Y, Tamura M, Ogata E, Matsumoto T. Stimulation of fracture repair by recombinant human basic fibroblast growth factor in normal and streptozotocin-diabetic rats. Endocrinology 1994;135:774–781.
25. Nielsen HM, Andreassen TT, Ledet T, Oxlund H. Local injection of TGFβ increases the strength of tibial fractures in the rat. Acta Orthop Scand 1994;65(1):37–41.
26. Lind M, Schumacker B, Soballe K, Keller J, Melsen F, Bunger C. Transforming growth factor β enhances fracture healing in rabbit tibiae. Acta Orthop Scand 1993;64(5):553–556.
27. Noda M, Camilliere JJ. In vivo stimulation of bone formation by transforming growth factor-β. Endocrinology 1989;124:2991–2994.
28. Taniguchi Y, Tanaka T, Gotoh K, Satoh R, Inazu M. Transforming growth factor-β1-induced cellular heterogeneity in the periosteum of rat parietal bones. Calcif Tis Int 1993;53:122–126.
29. Jingushi S, Bolander ME. Adverse effects of local administration of basic fibroblast growth factor and transforming growth factor β1 on fracture healing. Orthop Trans 1991;15(2):403–404.

Tissue Engineering for Therapeutic Use 1.
Y. Ikada and Y. Yamaoka, editors.

The use of natural metabolites in the design of new materials for tissue engineering

Joachim Kohn
Department of Chemistry, Rutgers — The State University of New Jersey, Piscataway, New Jersey, USA

Keywords: biocompatibility, biodegradation, biomaterials, polyarylates, polycarbonates, tyrosine.

Introduction

In 1988, an estimated 17 million Americans carried major medical implants in their bodies [1]. Improvements in the cost effectiveness and safety of medical implants, and the identification of new treatment options (such as tissue reconstruction) will translate into a further dramatic increase in the number of patients benefitting from medical implants [2].

A critical issue facing the emerging field of tissue engineering today is the lack of a broad range of materials for the manufacture of polymeric scaffolds and for the systematic study of cell-substrate interactions [3]. Metals and various industrial plastics, e.g., polysiloxanes, polyurethanes, Dacron®, Teflon®, polyethylene, are the most commonly used biomaterials. These biostable, synthetic implant materials lack the biological sequences and patterns crucial to normal cell function and often trigger aberrant cell responses. The fact that only very few degradable polymers are available to the biomaterials scientist and the tissue engineer represents an important and critical limitation. In the USA, for example, poly(glycolic acid), poly(lactic acid), polydioxanone and copolymers thereof are the only synthetic, degradable polymers with extensive FDA approval history.

Over the last decade, exciting, new applications for medical implants have been identified which all require degradable biomaterials. Example applications are polymeric scaffolds for tissue reconstruction, development of artificial organs, degradable membranes for the prevention of surgical adhesions, and intra-arterial coatings for the prevention of thrombosis or restenosis.

Most of these new applications are associated with the rapidly developing field of tissue engineering which holds the promise of solving many medical and surgical problems requiring the repair of damaged or diseased tissues. For example, the feasibility of creating tissue engineered bone was shown in two seminal works. Vacanti et al. seeded fibrous scaffolds of poly(glycolic acid) in vitro with allogenic

Address for correspondence: Prof Joachim Kohn, Rutgers University, Department of Chemistry, Piscataway, NJ 08855-0939, USA. Tel.: +1-908-445-3888. Fax: +1-908-445-5006.
E-mail: Kohn@rutchem.rutgers.edu

osteoblasts and then implanted the scaffolds subcutaneously in nude mice [4]. Later, Wakitani et al. [5] showed in rabbits that a convenient source of host-derived mesenchymal stem cells could be grown in culture and could be seeded on collagen scaffolds. After implantation in full-thickness cartilage defects, the cells retained their ability to differentiate into bone or cartilage depending on their location within the implant.

It is important to emphasize that the clinical implementation of these advanced biomaterials' concepts will depend in part on the availability of degradable biomaterials whose physicomechanical, chemical, and biological properties can be matched closely to a set of predetermined requirements. Indeed, one of the key scientific challenges [6–9] still to be addressed in tissue engineering is the lack of novel scaffold materials — new synthetic degradable polymers that specifically address the need for increased biocompatibility in the short and long term, and that allow one to tailor cellular interactions to promote cell specific proliferation and maintenance of differentiated function.

Other important research challenges in tissue engineering relate to the systematic study of tissue-specific responses to the pore architecture and morphology of polymeric scaffolds, and to the question of whether it is indeed advantageous to pre-seed scaffolds in vitro prior to in vivo implantation. However, these challenges are secondary to the limited variety of suitable, degradable polymers available to the tissue engineer. For this reason, this article is focused on the "materials base" of tissue engineering and several classes of new synthetic degradable polymers are described.

Synthetic degradable polymers used in tissue engineering

Tissue engineers have employed almost exclusively poly(glycolic acid) (PGA), poly(lactic acid) (PLA), and other poly(α-hydroxy ester)-based materials for the construction of polymeric scaffolds. Although the utility of these materials as sutures and in a number of drug delivery applications is well-established, a wide range of material needs cannot be addressed by the use of the above-mentioned polyesters. For example, all polyesters release acidic degradation products [10–12], limiting their utility to applications where acidity at the implant site is not a concern. The above polyesters tend to be relatively rigid, inflexible materials [13]. This can be an advantage in load-bearing applications, but is a definite disadvantage when mechanical compliance with soft tissue or blood vessels is required. Finally, none of the above polyesters provides a chemically reactive pendent chain for the easy attachment of drugs, cross-linkers, or biologically active moieties. Thus, the simple poly(α-hydroxy ester)-based materials have performed well in establishing the foundation and feasibility of the tissue engineering approach [4,6,7,14–16], but may not be the optimum type of material to use in the construction of polymeric scaffolds.

One reason for the predominant use of PGA, PLA and their copolymers by virtually every laboratory working on polymeric scaffolds for tissue engineering, is the easy availability of these polymers in medical grade from a number of commercial sources. A second reason is the assumption that it will be easier to obtain regulatory approval for the commercial sale of any tissue engineering product if only known polymers such as PGA or PLA are used. Thirdly, there are already a wide range of reliable procedures described in the literature for the fabrication of polymeric scaffolds — almost all of these procedures use PGA/PLA or copolymers thereof.

For example, degradable polymeric scaffolds are routinely fabricated using leaching techniques where a porogen, e.g., sodium chloride crystals, is removed from a solvent-casted polymer/porogen composite [7]. The commonly used sodium chloride crystals result in open pore scaffolds with pore sizes that can range from 50 to 500 μm with quite narrow distributions. However, the pores have a harsh, cubic structure. More recently, phase separation and foaming techniques from industry have been successfully adapted for use in the fabrication of polymeric scaffolds [17,18]. Highly porous structures are produced with, in some cases, spherical pores, but the pores are typically much smaller, ranging in size from submicron to several microns. In spite of a large number of publications, improved fabrication methods are still needed that will allow systematic alterations in pore shape, pore interconnectivity, and pore size distributions.

New degradable polymers derived from the natural amino acid L-tyrosine

Amino-acid-derived polymers can be conveniently classified into conventional poly(amino acids) and pseudo-poly(amino acids). Conventional poly(amino acids) are synthetic polymers composed of α-amino acids linked by peptide bonds while pseudo-poly(amino acids) are synthetic polymers composed of α-amino acids and other components linked by nonpeptide bonds, such as ester or carbonate linkages. The term pseudo-poly(amino acid) is used in analogy to the term pseudopeptide which describes a peptide containing nonamide backbone linkages [19]. Among various pseudo-poly(amino acids) [20,21], tyrosine-derived polycarbonates have been investigated extensively. The exact chemical structure of these materials is shown in Fig. 1. Several detailed reviews are available describing the material properties of these polymers in more detail [21—24].

Physicomechanical properties

Tyrosine-derived polycarbonates are fully amorphous polymers, characterized by their relatively high strength and stiffness (Table 1), which exceeds that of other degradable polymers such as poly(ε-caprolactone) and poly(ortho esters) but is somewhat lower than the values reported for poly(L-lactic acid) (2.4—10 GPa) and poly(glycolic acid) (6.5 GPa) [13,25]. The mechanical properties of tyrosine-derived polycarbonates make them amenable to low load-bearing devices such as would be required in small bone fixation.

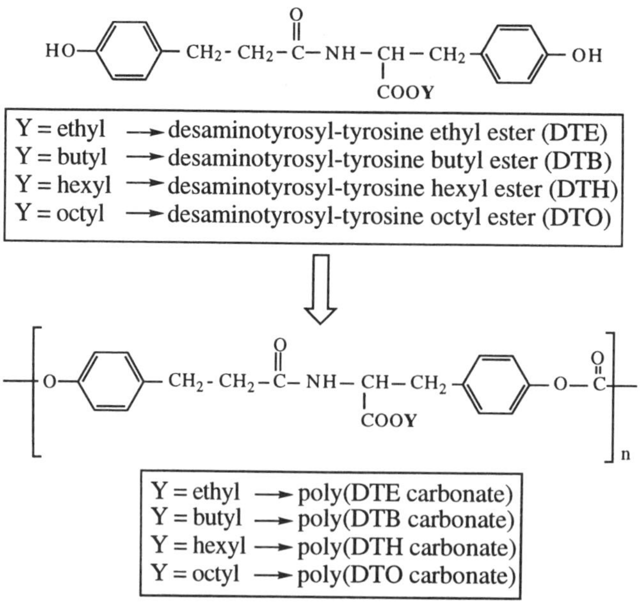

Fig. 1. Chemical structure and nomenclature for four tyrosine-derived polycarbonates. Note that the polymers form a homologous series of materials differing only in the length of their respective pendent chain. The monomers are derivatives of tyrosine-dipeptide where the amino terminus has been replaced by a hydrogen atom. These monomers can be polymerized using phosgene using the same basic procedures employed commercially for the preparation of industrial polycarbonates. (Reproduced from [29].)

Degradation mechanism

Generally, tyrosine-derived polycarbonates degrade relatively slowly, on the order of months to years [22]. The degradation mechanism of tyrosine-derived polycarbonates has been carefully investigated combining surface analysis (ESCA, ATR-FTIR), gel permeation chromatography, kinetic modeling, and the identification of degradation products by HPLC. These studies led to the postulation of an in vitro mechanism of degradation (Fig. 2) [22,26].

Table 1. Physical properties of tyrosine-derived polycarbonates[a].

Polymer derived from[a]	M_w[a]	Tg (°C)	Contact angle (°)	Young's modulus (GPa)	Tensile strength (MPa)
DTE	176000	81	73	1.5	67
DTB	120000	66	77	1.6	60
DTH	350000	58	86	1.4	62
DTO	450000	53	90	1.2	51

[a]Data from Ertel and Kohn [22]. Weight average molecular weights as determined by GPC.

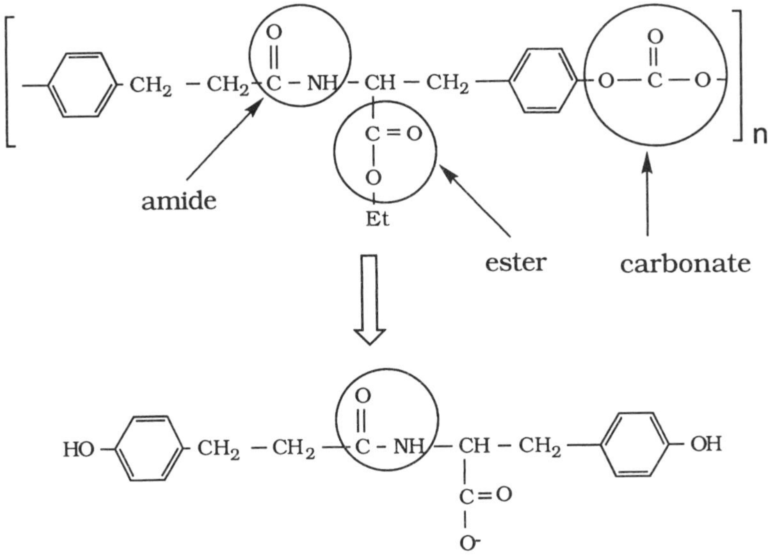

Fig. 2. The degradation mechanism in vitro involves the concomitant cleavage of the ester pendent chain and the backbone carbonate bonds. Although overall water uptake is low (up to 5 vs. >20% for PLA), the degradation of the backbone occurs throughout the bulk of the polymer. This mechanism predicts the formation of desaminotyrosyl-tyrosine as the major degradation product. This was confirmed experimentally. The approximate half-life for the cleavage of the polymer backbone is about 25–50 weeks and depends on the device configuration (rods vs. thin films) and also on the nature of the pendent chain. In vivo, the further cleavage of the amide bond may lead to the formation of desaminotyrosine and tyrosine as the final degradation products. (Reproduced from [29].)

Processibility

Since tyrosine-derived polycarbonates are completely amorphous, they are easy to process into films, fibers, microspheres, and membranes by conventional solvent casting and evaporation techniques [13,20,27]. In addition, the wide gap between a relatively low glass transition temperature (below 100 °C) and a relatively high decomposition temperature (above 300 °C) makes tyrosine-derived polycarbonates easy to extrude, compression mold, or fabricate into complex shapes by injection molding. Compression molded films, extruded rods and films, and drawn fibers have been obtained by the authors. In addition, being good film formers, tyrosine-derived polycarbonates are readily amenable to the fabrication of porous scaffolds by salt leaching or phase separation techniques. Porous scaffolds with promising pore architectures have been described in the literature [28].

Sterilization of tyrosine-derived polycarbonates

In order for tyrosine-derived polycarbonates to be utilized as medical devices, they must be able to withstand the conditions associated with sterilization. A key issue for the preparation of porous scaffolds is the inability of PLA and PGA to withstand sterilization by γ-irradiation, while ethylene oxide treatment of porous scaffolds may require excessive degassing times to remove all traces of ethylene oxide from the interior of the scaffold. In a recent study [29], solvent cast films of tyrosine-derived polycarbonates were exposed to γ-irradiation (ranging from 0.3 Mrad to 10 Mrad). Molecular weight retention, mechanical properties, surface chemical changes, and lingering effects on postirradiation behavior of the γ-sterilized materials were determined. Poly(L-lactic acid) was included as a control. With γ-irradiation at a clinically relevant dose of 3.3 Mrad, the polycarbonates retained 80% or more of their initial molecular weight. In correspondence with known data [30,31], poly(L-lactic acid) proved to be much more sensitive to γ-irradiation and at 3.3 Mrad retained only 50% of its original molecular weight.

Biocompatibility of tyrosine-derived polycarbonates in comparison with polydioxanone and PLA

Bone repair and fixation via degradable, polymeric implants is a promising alternative to metallic devices. Currently, pins made of PGA and polydioxanone are in clinical use in the USA, while PLA has been widely investigated but not yet approved as a small bone fixation device in the USA [32–34]. A number of publications linked the acidic nature of the degradation products and the release of crystalline particulate debris to a documented late inflammatory and bone resorption response [34–37]. Unlike PLA, tyrosine-derived polycarbonates are amorphous and will not degrade to crystalline particulates nor are the degradation products significantly acidic. Indeed, in several studies, the biocompatibility of tyrosine-derived polycarbonates was evaluated in side-by-side comparisons with medical grade poly(L-lactic acid).

An assay of biocompatibility in a subcutaneous rat model showed that the soft-tissue response to poly(DTH carbonate) is not significantly different from the responses elicited by medical grade PLA or polyethylene [38]. This initial study was followed by a detailed evaluation of the bone response to poly(DTH carbonate) in a side-by-side comparison with polydioxanone in a rabbit model [23] and by a comparative study of poly(DTE carbonate), poly(DTH carbonate), and medical grade PLA in a canine bone chamber model [24]. Key conclusions of these studies are summarized below.

Results and conclusions of the rabbit model

Pins of poly(DTH carbonate) and polydioxanone (known as Orthosorb or PDS) were implanted transcortically in the distal femur and proximal tibia of New Zealand white rabbits, cell growth was labeled by fluorescent dyes, and the implant sites were retrieved for histological analysis at 1, 2, 4 and 26 weeks.

Throughout the 26-week experimental period, poly(DTH carbonate) implants were in close contact with bone with very little intervening fibrous tissue layer and there were no signs of an inflammatory response or bone resorption. At 26 weeks, the implant presented a rough surface and surface crevices were filled with new bone and fibrous tissue. Under UV light, the implants were surrounded by yellow fluorescent rims suggesting intense osteoblastic activity around 13 weeks and, possibly, a slower bone growth process afterward. Backscattered electron images indicated that the bone surrounding the implant had evolved into organized lamellar bone which was free of marrow spaces. The images indicated that the bony protuberances which penetrated the implant were calcified. The formation of calcified bone into a degrading implant had never been observed for polyester-based implants.

Significantly different findings were observed for polydioxanone. For this polymer, close bone contact was observed less frequently and the implants were typically separated from the bone by a 25- to 100-μm thick fibrous layer consisting of fibroblast-like cells. Osteoclastic cavities were observed close to the implant. At 26 weeks, the six explanted polydioxanone pins exhibited a variety of responses which may be linked to the degree of degradation reached by the polydioxanone pins. For one pin, no sign of the implanted polymer could be found in the cross sections. A total of two pins were completely resorbed and only a small layer of scar tissue marked the initial periphery of the pins. The remaining three pins showed various degrees of inflammation that ranged from mild to severe and were characterized by loose granulation tissue replacing the implants, along with tissue necrosis at the periphery. No fluorescent markers had been incorporated in the tissue surrounding the implant. Under polarized light, highly crystalline debris of polydioxanone could be detected. The debris were incorporated in the bone matrix and had migrated away from the implant site.

This study provided the first indication of two noteworthy differences in the bone response between polyesters and tyrosine-derived polycarbonates: 1) the degree of direct contact between the bone and the implant, and 2) the gradual replacement of the polycarbonate material by functional bone tissue for tyrosine-derived polycarbonates as opposed to a fast and disruptive resorption process for polydioxanone.

Results and conclusions of the canine bone chamber model

The canine bone chamber model [34,39,40] employs an implant placed in a surgically created cortical defect in the distal femur consisting of 10 channels lined by the test polymer. By measuring bone ingrowth into the channels and examining histological sections of the bone-implant interface, an assessment of bone biocompatibility can be made. Poly(DTE carbonate), poly(DTH carbonate) and medical grade PLA were compared. Bone ingrowth into the chambers was assessed at 6, 12, 24 and 48 weeks.

Bone ingrowth was evident as early as 6 weeks for each of the polymer test chambers and continued to progress for the first 24 weeks. The 24-week implants showed that the poly(DTE carbonate) channels were completely filled with bone. Such strong osteoconductivity was not seen in any of the PLA chambers. In fact, the bone chamber model had been used before to screen a number of materials, including metals, and complete filling of all channels had previously only been observed for hydroxyapatite [24]. This difference in the biological response to tyrosine-derived polycarbonates and PLA became even more striking at 48 weeks. Whereas bone ingrowth into the PLA chambers decreased to less than half of the 24 week ingrowth average, bone ingrowth into the poly(DTE carbonate) and poly(DTH carbonate) chambers was sustained.

Histological analysis of the bone-implant interface also revealed a fundamental difference in the response to tyrosine-derived polycarbonates and PLA. Consistent in all PLA chambers examined was the presence of an intervening fibrous tissue layer between the bone and test material preventing direct bone apposition to the PLA surface. This consistent fibrous layer was notably absent in the histologic sections of poly(DTE carbonate) and poly(DTH carbonate) chambers where direct bone apposition was evident. Color histological micrographs have been published [24].

Summary

The synthesis and characterization of tyrosine-derived polycarbonates has been reported in the literature. Currently, four polymers, differing only in the length of their pendent chain (ethyl, butyl, hexyl and octyl) are available. Among these polymers, poly(DTE carbonate) was found to be most suitable for use in orthopaedic implants due to its physicomechanical properties, its strength retention profile, and its ability to support the attachment and growth of hard-tissue-related cells in vitro and in vivo. For this polymer, clinical trials in humans may commence in the near future. The biological response of hard tissue to tyrosine-derived polycarbonates appears to be fundamentally different from the response observed for PLA and polydioxanone. Particularly relevant is the observation that poly(DTE carbonate) is strongly osteoconductive in the canine model. The data given here indicate that advantages can be realized by employing novel polymers (instead of PLA or PGA) as medical implant materials in selected applications. The use of tyrosine-derived polycarbonates in the design of polymeric scaffold for hard-tissue reconstruction, bone graft substitutes, and spine fusion is currently under investigation.

Acknowledgements

This work was a collaborative research project of the New Jersey Center for Biomaterials and Medical Devices. The collaboration of Drs M. Zimmerman, R. Parsons, H. Alexander and J. Ricci is gratefully acknowledged. In addition, Drs Kenneth James and Sylvie I. Ertel, Mr Varawut Tangpasuthadol and Ms Kimberly A. Hooper contributed to the work described here. Tyrosine-derived polycarbonates were first prepared under support from Zimmer Inc. (Warsaw, Indiana, USA). Further work was funded by NIH grant GM39455, and by Integra Lifesciences Inc. (Plainfield, New Jersey, USA).

References

1. Moss AJ, Hamburger S, Jeng LL, Howie LJ. Use of selected medical device implants in the United States, 1988. Advance Data from Vital and Health Statistics of the National Center for Health Statistics, vol 191. Washington, DC: US Department of Health and Human Services, 1991.
2. Kohn J. Biomaterials science at a crossroads: are current product liability laws in the United States hampering innovation and the development of safer medical implants. Pharmacol Res 1996;13(6):815–819.
3. Aronoff Associates, Biomaterials Crisis: Market Study: Biomaterials Supply for Permanent Medical Implants. Report prepared for the Health Industry Manufacturers Association by Aronoff Associates, New York, NY, March 1994.
4. Vacanti CA, Kim W, Upton J, Vacanti MP, Mooney D, Schloo B, Vacanti JP. Tissue engineered growth of bone and cartilage. Transpl Proc 1993;25:1019–1021.
5. Wakitani S, Goto T, Pineda SJ, Young RG, Mansour JP, Kaplan AI, Goldberg VM. Mesenchymal cell-based repair of large, full-thickness defects of articular cartilage. J Bone Joint Surg 1994;76A:579–592.
6. Thomson RC, Wake MC, Yaszemski MJ, Mikos AG. Biodegradable polymer scaffolds to regenerate organs. Adv Polymer Sci 1995;122:245–274.
7. Mikos AG, Sarakins G, Leite SM, Vacanti JP, Langer R. Laminated three-dimensional biodegradable foams for use in tissue engineering. Biomaterials 1993;14:323–329.
8. Mikos AG, Sarakinos G, Lyman MD, Ingber DE, Vacanti JP, Langer R. Prevascularization of porous biodegradable polymers. Biotechnol Bioeng 1993;42:716–723.
9. Mooney DJ, Organ G, Vacanti JP, Langer R. Design and fabrication of biodegradable polymer devices to engineer tubular structures. Cell Transpl 1994;3:203–210.
10. Böstman O, Hirvensalo E, Vainionpää S, Mäkelä A, Vihtonen K, Törmälä P, Rokkanen P. Ankle fractures treated using biodegradable internal fixation. Clin Orthop 1989;238:195–203.
11. Vert M, Christel P, Chabot F, Leray J. Bioresorbable plastic materials for bone surgery. In: Hastings GW, Ducheyne P (eds) Macromolecular Biomaterials. Boca Raton: CRC Press, Inc., 1984;119–142.
12. Vert M, Li S, Garreau H. More about the degradation of LA/GA-derived matrices in aqueous media. J Contr Res 1991;16:15–26.
13. Engelberg I, Kohn J. Physico-mechanical properties of degradable polymers used in medical applications: a comparative study. Biomaterials 1991;12(3):292–304.
14. Mikos AG, Bao Y, Cima LG, Ingber DE, Vacanti JP, Langer R. Preparation of poly(glycolic acid) bonded fiber structures for cell attachment and transplantation. J Biomed Mater Res 199327:183–189.
15. Vacanti CA, Langer R, Schloo B, Vacanti JP. Synthetic polymers seeded with chondrocytes provide a template for new cartilage formation. Plast Recon Surg 1991;88(5):753–759.
16. Freed LE, Marquis JC, Nohria A, Emmanual J, Mikos AG, Langer R. Neocartilage formation in vitro and in vivo using cells cultured on synthetic biodegradable polymers. J Biomed Mater Res 1993;27:11–23.
17. de Groot JH, Nijenhuis AJ, Bruin P, Pennings AJ, Veth RPH, Klompmaker J. Use of porous biodegradable polymer implants in meniscus reconstruction. 1) Preparation of porous biodegradable polyurethanes for the reconstruction of meniscus lesions. Coll Polym Sci 1990;268: 1073–1081.
18. Schugens C, Maquet V, Grandfils C, Jerome R, Teyssie P. Polylactide macroporous biodegradable implants for cell transplantation. II. Preparation of polylactide foams by liquid-liquid phase separation. J Biomed Mater Res 1996;30:449–462.
19. Spatola AF. Peptide backbone modifications: a structure-activity analysis of peptides containing amide bond surrogates, conformational constraints and related backbone replacements. In: Weinstein B (ed) Chemistry and Biochemistry of Amino Acids, Peptides, and Proteins. New York, NY: Marcel Dekker, 1983;267–357.

20. Kohn J. Design, synthesis, and possible applications of pseudo-poly(amino acids). Trends Polym Sci 1993;1(7):206–212.
21. James K, Kohn J. Applications of pseudo-poly(amino acid) biomaterials. Trends Polym Sci 1996;4(12):394–397.
22. Ertel SI, Kohn J. Evaluation of a series of tyrosine-derived polycarbonates for biomaterial applications. J Biomed Mater Res 1994;28:919–930.
23. Ertel SI, Kohn J, Zimmerman MC, Parsons JR. Evaluation of poly(DTH carbonate), a tyrosine-derived degradable polymer, for orthopaedic applications. J Biomed Mater Res 1995;29(11): 1337–1348.
24. Choueka J, Charvet JL, Koval KJ, Alexander H, James KS, Hooper KA, Kohn J. Canine bone response to tyrosine-derived polycarbonates and poly(L-lactic acid). J Biomed Mater Res 1996; 31:35–41.
25. Daniels AU, Chang MKO, Andriano KP, Heller J. Mechanical properties of biodegradable polymers and composites proposed for internal fixation of bone. J Appl Biomater 1990;1:57–78.
26. Ertel SI, Kohn J. A Homologous Series of Tyrosine-Derived Polycarbonates: Surface Characterization and Degradation. The 19th Annual Meeting of the Society of Biomaterials, Birmingham, Alabama, USA, 1993.
27. Kohn J, Yu C, Yeo SD, Debenedetti PG. Microspheres From Tyrosine-Based Polymers: Evaluation of Processing Parameters and Initial Release Data. Proceedings of the 21st International Symposium on Controlled Release of Bioactive Materials, Nice (France), 1994.
28. James K, Kohn J. New biomaterials for tissue engineering. MRS Bull 1996;21(11):22–26.
29. Hooper KA, Cox JD, Kohn J. Comparison of the effect of ethylene oxide and γ-irradiation on selected tyrosine-derived polycarbonates and poly(L-lactic acid). J Appl Polym Sci 1997; 63(11):1499–1510.
30. Martakis N, Niaounakis M, Pissimissis D. γ-Sterilization effects and influence of the molecular weight distribution on the postirradiation resistance of polypropylene for medical devices. J Appl Polym Sci 1994;51:313–328.
31. Shalaby SW, Johnson RA. Synthetic absorbable polyesters. In: Shalaby SW (ed) Biomedical Polymers: Designed-to-Degrade Systems. Munich: Hanser Publishers, 1994;22,27.
32. Pistner H, Gutwald R, Ordung R, Reuther J, Mühling J. Poly(L-lactide): a long-term degradation study in vivo. I. Biological results. Biomaterials 1993;14(9):671–677.
33. Bucholz RW, Henry S, Henley MB. Fixation with bioabsorbable screws for the treatment of fractures of the ankle. J Bone Joint Surg Am 1994;76(3):319–324.
34. Suganuma J, Alexander H. Biological response of intramedullary bone to poly-L-lactic acid. J Appl Biomater 1993;4:13–27.
35. Böstman OM. Absorbable implants for the fixation of fractures. J Bone Joint Surg 1991;73(1): 148–153.
36. Daniels AU, Taylor MS, Andriano KP, Heller J. Toxicity of absorbable polymers proposed for fracture fixation devices. Proc Orthop Res Soc 1992;17:88.
37. Taylor MS, Daniels AU, Andriano KP, Heller J. Six bioabsorbable polymers: in vitro acute toxicity of accumulated degradation products. J Appl Biomater 1994;5:151–157.
38. Silver FH, Marks M, Kato YP, Li C, Pulapura S, Kohn J. Tissue compatibility of tyrosine-derived polycarbonates and polyiminocarbonates: an initial evaluation. J Long-Term Eff Med Implants 1992;1(4):329–346.
39. Ricci JL, Spivak JM, Blumenthal NC, Alexander H. Modulation of bone ingrowth by surface chemistry and roughness. In: Davies J (ed) The Bone-Biomaterial Interface. Toronto, Buffalo, London: University of Toronto Press, 1991;334–349.
40. Spivak JM, Blumenthal NC, Ricci JL, Alexander H. A new canine model to evaluate the biological effects of implant materials and surface coatings on intramedullary bone ingrowth. Biomaterials 1990;11(1):79–82.

Tissue Engineering for Therapeutic Use 1.
Y. Ikada and Y. Yamaoka, editors.

Present status of an artificial liver support system for hepatic failure

Shinichi Kasai and Masayuki Sawa
Second Department of Surgery, Asahikawa Medical College, Asahikawa, Japan

Keywords: artificial liver, extracorporeal perfusion, hepatic failure, hepatocytes.

Introduction

The liver is an important organ which plays a key role in metabolism, and severe hepatic insufficiency has a lethal outcome. The therapeutic modalities for this lethal pathophysiological entity are artificial liver support (ALS) and liver transplantation. The former involves hemodiafiltration (HDF), hemoadsorption, and plasmapheresis, but the outcome is still unsatisfactory [1]. Liver transplantation from brain-dead donors is performed to replace irreversibly damaged livers and has become a treatment of choice for patients with acute as well as chronic hepatic failure [2]. The overall 1-year survival rate after liver transplantation is over 80% and that of patients with fulminant hepatic failure (FHF) is reported to be approximately 60%. The most difficult issue in liver transplantation is the shortage of donor organs as approximately 30% of patients on waiting lists die before a donor liver becomes available. The development of a clinically applicable ALS system would thus be desirable, but this has not yet been possible due to the difficulty in reproducing the complicated functions of the liver. In this article, we first review the history of ALS systems, and then discuss recent clinical approaches and ongoing basic research including our own studies.

History of ALS systems

Figure 1 summarizes the history of ALS systems. In 1914, Abel et al. attempted to remove toxic substances from the blood by dialysis using a collodion membrane [3]. In 1956, Killey et al. applied a cellophane membrane hemodialyser for the treatment of hepatic coma [4]. In addition, Krebs et al. focused on liver enzymes and found that sliced liver tissue could metabolize ammonia to urea nitrogen [5].

In the 1950s, the idea of ALS was established under the influence of the devel-

Address for correspondence: Shinichi Kasai MD PhD, Second Department of Surgery, Asahikawa Medical College, 4-5 Nishikagura, Asahikawa 078-8510, Japan. Tel.: +81-166-68-2500. Fax: +81-166-68-2193. E-mail: kasai@asahikawa-med.ac.jp

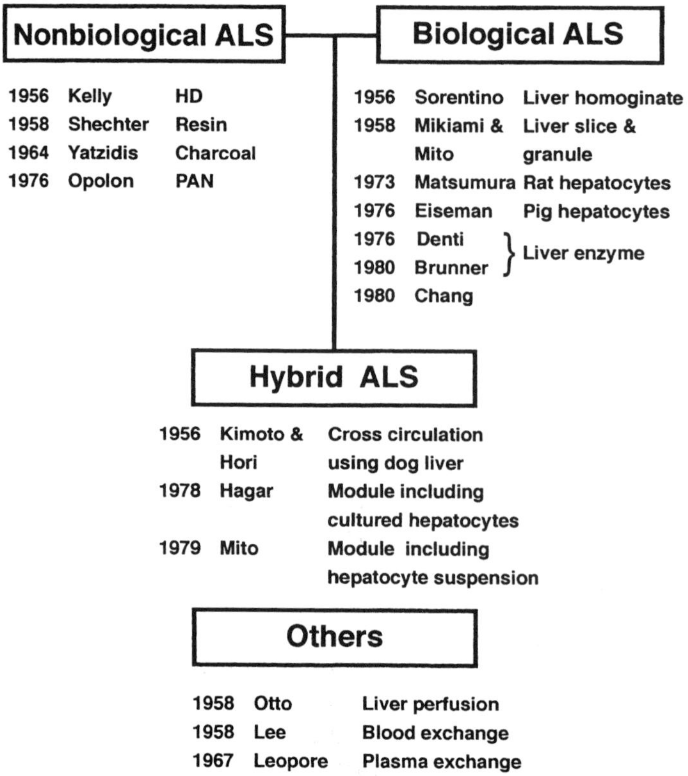

Fig. 1. Historic review of artificial liver support systems in early era.

opment of clinically applicable artificial kidneys. In 1957, Kimoto et al. developed a device for dialysis of human and dog blood through a dialysis membrane and the device was named an "artificial liver" for the first time [6]. Sorrentino et al. found that a device containing bovine liver homogenate could metabolize salicylic acid and barbital in the blood, and could synthesize urea nitrogen from ammonia. They reported on this device as "fegatto artificiale" [7]. Mikami et al. developed a device containing sliced liver tissue or freeze-dried liver granules and performed clinical trials [8].

Schechter focused on the ability of an ion-exchange resin to remove ammonia and succeeded in clinical trials of extracorporeal perfusion with an ion-exchange resin column [9]. Studies on adsorbents led to the development of charcoal by Yatzidis in 1964 [10].

Almost all the ideas for ALS systems originated in this era, but no widely applicable ALS system was developed, and clinical trials of blood exchange or extracorporeal liver perfusion were carried out at some limited institutions. In 1974, the group of Williams in the UK reported revolutionary results (survival rate: 45%) with charcoal hemoperfusion [11]. In those days, artificial kidneys were already an established therapeutic modality for renal failure. Various kinds of membranes with a high antithrombogenecity were available due to marked advances in polymers for medicine and medical electronics. On the basis of these technological advances, charcoal hemoadsorption, polyacrylnitrile (PAN) membrane dialysis, and plasma exchange were clinically applied for the treatment of patients with hepatic failure. The present status of these therapeutic modalities will be discussed in more detail later.

Pathophysiology of hepatic failure

Hepatic failure is the state caused by massive necrosis of liver parenchymal cells and subsequent insufficiency of liver function. Hepatic failure causes various neurological symptoms and is classified as acute or chronic. Acute hepatic failure (AHF) is caused by viral hepatitis, drug overdose, and narcotic agents. Patients rapidly fall into a coma. In contrast, chronic hepatic failure (CHF) is usually found in patients with end-stage liver cirrhosis. There are thus two different pathophysiologies of hepatic failure, the main target of ALS should be the acute type, which is defined as fulminant hepatitis.

The most distinctive feature of hepatic failure is the disturbance of consciousness and there are some patients who fall into deep coma within just a few hours. Once in coma, the mortality rate is closely related to the severity of brain damage [12], and thus attempts to reawaken the patient should be made very quickly. The causes of hepatic coma are thought to be lack of hepatic factors to maintain the central nervous system due to hepatic functional insufficiency [13], accumulation of toxic substances [14], or both of the above [15]. However, no definite theory has been established.

Ammonia, free fatty acids, mercaptan, phenol, bilirubin, bile acids and amines have been reported as toxic substances which may be responsible for hepatic coma. Toxicity of each substance has been shown in animal experiments and clinical studies, but no direct correlation with the onset of hepatic coma has been confirmed. Hughes et al. found an abnormal peak at a molecular weight of less than 1,000 by analyzing ultrafiltrated human serum using Sephadex G25 chromatography and suggested that this fraction depressed Na·K-ATPase and was one of the causes of cerebral edema [16]. Chang, Inoue, and Opolon also reported an abnormal peak at a molecular weight of 500–5,000 that was removed by charcoal hemoadsorption [17–19]. Further investigations are required to determine whether these are the same substance and whether it is biologically active.

Present status and problems in the treatment of fulminant hepatitis

The basic treatment of fulminant hepatitis is as follows: 1) rapid reversal of hepatic coma by removing toxic substances, supplying essential substances, and lowering the elevated intracranial pressure; 2) providing the conditions necessary for liver regeneration by preventing further hepatocellular necrosis; and 3) prevention of lethal complications. In this article, the present status and problems of clinically applied ALS systems, including plasma exchange and extracorporeal liver perfusion, are discussed (Fig. 2).

Hemoperfusion

Hemoperfusion is performed by bringing the blood into direct contact with adsorbents such as charcoal and ion-exchange resins that adsorb and remove toxic substances from the blood. This procedure was first developed by Yatzidis in 1964 using coconut charcoal [10]. Since the blood comes into direct contact with the adsorbents, there is a danger that blood cell components will also be adsorbed, along with the danger of particle release into the blood. Therefore, new adsorbents that do not release particles are now being developed and various coating materials are also employed.

The absorption range of charcoal is between a molecular weight of 300 and 1,500, and is effective for water-soluble and non-protein-bound substances. Gimson et al. [20] have reported that administration of prostacyclin (PG-I_2) is effective in controlling hypotension resulting from platelet aggregation during hemoperfusion.

Ion-exchange resins have also been studied as adsorbents. Various resins can be produced with different electrical charges, including neutral, cationic, and anionic exchange resins. Presently a neutral ion exchange resin is being utilized clinically and it shows outstanding adsorption of lipophilic and protein-bound substances.

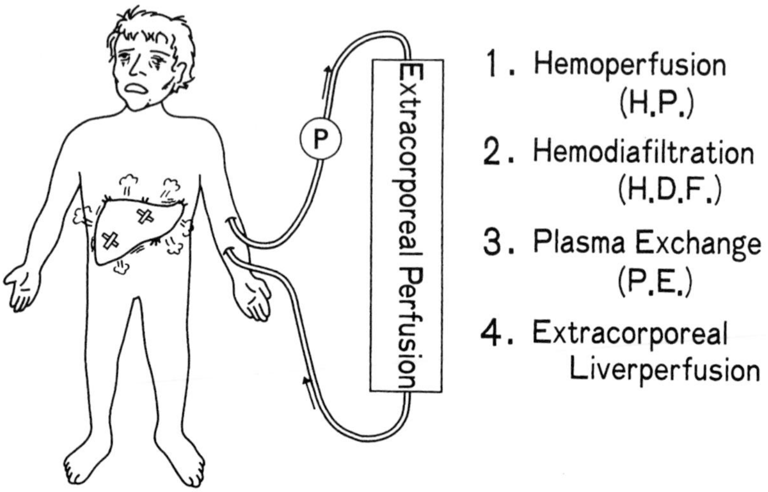

Fig. 2. Clinically applied artificial liver support systems.

Hemodialysis and hemofiltration

Hemodialysis was first used by Killey et al. [4] for the treatment of hyperammonemia due to liver disease. Since the dialysis method only removed substances with a molecular weight of less than 300, Opolon et al. [21] developed a PAN membrane with a large pore size. The PAN membrane has a dialysis range up to a molecular weight of ~5,000. To improve the removal rate of substances larger than this, hemofiltration has recently been combined with dialysis. The hemofiltration system has no detrimental effect on blood cell components, but its capacity for removing lipophilic substances and protein-bound substances such as bilirubin is nil.

Plasmapheresis

Plasmapheresis involves firstly separating the blood into cells and plasma. The separated plasma is then discarded and replaced with fresh frozen plasma. This procedure is much more efficient in removing toxic substances than blood exchange, and clotting factors can be supplemented.

Recently, Yoshiba et al. [22] developed a new ALS system consisting of plasma exchange in combination with hemodiafiltration using high-performance membranes. This system allows the supply of essential plasma components from fresh frozen plasma and removes middle molecular weight toxins by HDF. According to their report, 90% of 25 patients with FHF regained consciousness and the survival rate was 55%. Further successful results are expected.

Whole-liver perfusion

In 1958, Otto [23] performed ex vivo perfusion with an extracted liver for liver support. Recently, there have only been two reports concerning whole-liver perfusion: one from West Germany in which direct perfusion was performed using a baboon liver, and the other report by Ozawa et al. [24] in which cross-hemodialysis was performed using a pig liver. In the case of baboons, the blood types were relatively similar to those of humans. Although antibodies against the baboon liver appeared, it was said that serious reactions such as anaphylaxis were not observed. With the pig liver, however, it was necessary to utilize a separation membrane between the patient's blood and the liver tissue. Using this system, an improvement in the energy status of the damaged liver was observed.

Evaluation of clinical results

A comparison of treatment results is difficult because many factors affect the prognosis. The general results are summarized in Table 1. With general supportive therapy, the survival rate is $< 20\%$. With PAN membrane dialysis and plasma exchange, the survival rate increases to ~30%. Whole-liver perfusion achieves

Table 1. Outcome of patients with FHF.

	Cases	Survival rate (%)	Reports
Conservative	232	6	Ohta (1972)
	53	15	Silk (1977)
	108	20	Opolon (1977)
PAN HD	39	23	Opolon (1977)
	108	29	Gimson (1982)
	43	19	Yamamoto (1985)
Charcoal HP	45	20	Gimson (1985)
	33	18	Amano (1985)
	57	9	J.N.S. (1986)
Plasma exchange	24	17	Amano (1985)
	15	33	Inoue (1985)
	27	33	Hirasawa (1985)
	280	25	J.N.S. (1986)
Liver perfusion	24	33	Germany (1980)

an extremely high rate of recovery of consciousness. This suggests the presence of toxic substances that provoke coma, and demonstrates the effectiveness of such treatments for removing these toxic substances. However, it is also evident that synchronous treatment to prevent the progression of liver cell necrosis, to normalize the in vivo environment, and to promote liver regeneration are necessary to improve the survival rate.

Hybrid artificial liver

Recently, a hybrid artificial liver utilizing the metabolic activity of biolivers has also been studied. Hybrid artificial livers aim at compensating for the defects of biological and artificial materials by combining the two systems. As a biological material, isolated hepatocytes have been investigated with respect to handling, preservation, and re-expression of cellular functions. The principle of the system is to metabolize toxic substances through a semipermeable membrane and synthesize essential substances for maintaining life and supporting the impaired host liver. Use of a semipermeable membrane avoids immunological reactions. For these purposes, many different sources and methods for hepatocyte immobilization have been studied (Table 1).

Hepatocyte suspension system

These systems enable substance exchange between hepatocytes suspended in culture medium and patient's blood or plasma through a semipermeable membrane. A sheet and a capillary hollow fiber membrane have been developed. In 1987, Matsumura et al. [25] attempted hemodialysis of a patient with inoperable malignant obstructive jaundice using a kiil-type dialyzer (Fig. 3) filled with isolated rabbit hepatocytes and reported a 68% reduction of jaundice along with symptomatic improvement.

We have also investigated the metabolic activity of isolated dog hepatocytes packed into a metabolic module with polymethylmethacrylate (PMMA) hollow fiber membranes (Fig. 3) [26,27]. Approximately 40 g of hepatocytes (8×10^9 cells), which correspond to 20–30% of the total number of hepatocytes in the liver of a 10 kg animal, were suspended in Eagle's minimum essential medium (MEM) and PMMA hollow fiber membranes were placed in the chamber. The molecular cutoff of the membranes was 40,000 or 100,000 Da, and the effective surface area was 0.4 m^2. The chamber was placed in a water bath heated at 37 °C. For ex vivo experiments, adult beagle dogs weighing about 10 kg were injected intravenously with 0.8 g/kg D-galactosamine (D-gal). Separated plasma was perfused through the device for 6 h. In animal experiments, serum ammonia, urea nitrogen and glucose levels were maintained within the normal range. The survival time of the dogs was as follows. In the nonsupported group, animals died 44 h after the D-gal injection. In the pseudoperfusion group without hepatocytes, the survival time was about 50 h. In the perfusion group with hepatocytes using a 100,000-Da membrane, survival time was over 80 h (Fig. 4). To observe the immunological effect of the metabolic support system, an ex vivo perfusion experiment was performed with a pig as the patient and a dog hepatocyte suspension. Although there were no immunological problems detected, the patients may have produced antibodies against dog hepatocytes. This suggests that further experiments are needed.

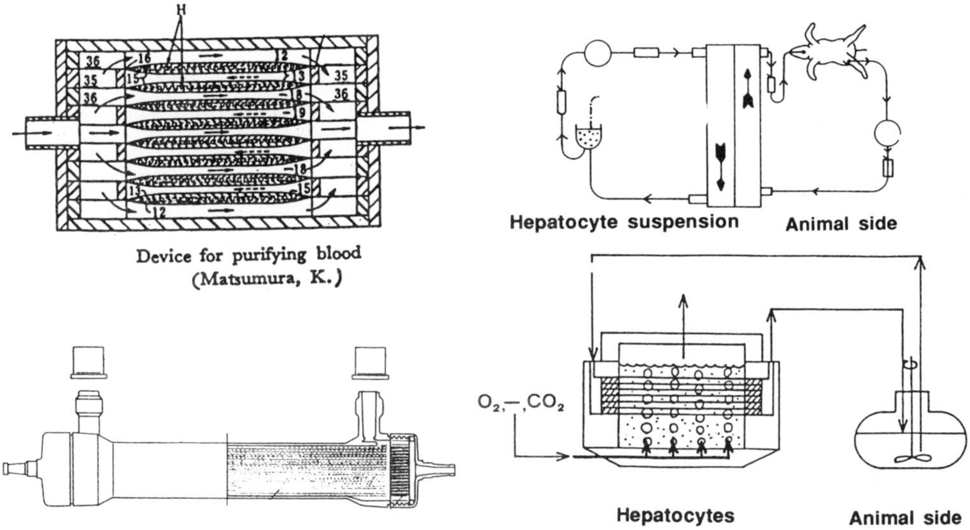

Fig. 3. Hybrid artificial livers using hepatocyte suspension system. Upper left: Matsumura et al.; upper right: Eiseman et al.; lower left and right: Kasai et al.

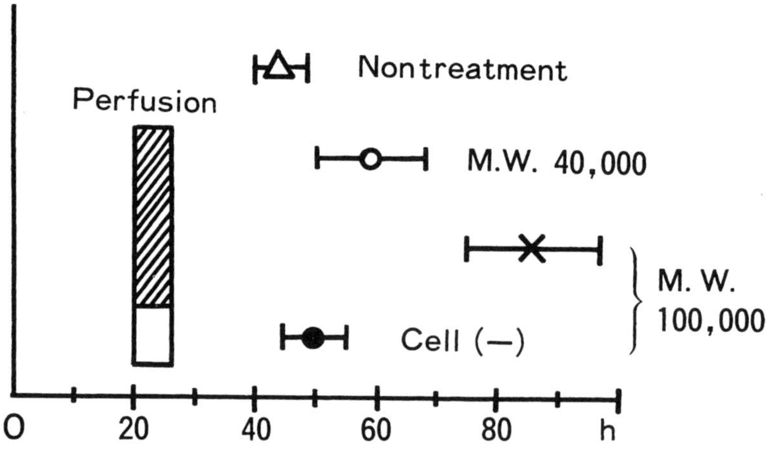

Fig. 4. The survival time of FHF dogs.

Recent experiments have shown that hepatocytes usually need to be attached to a substrate to maintain their function, so we examined the efficiency of many substrates for the maintenance of hepatocellular function. Collagen, poly-N-p-vinylbenzyl-D-lactonamide (PVLA), and a biomatrix prepared from dog livers all shortened the time necessary for cell attachment to culture dishes, and the metabolic activity of hepatocytes was increased by using microcarriers, biomatrix, and calcium alginate gel.

Isolated hepatocyte monolayers

Recent advances in cell culture techniques have introduced new media and matrices for hepatocyte culture and have facilitated long-term maintenance of cell viability and metabolic activity. Collagen is a well-known biomatrix that has been investigated for its applicability to the long-term culture of isolated hepatocytes. Hase et al. [28] developed a module containing 6×10^9 dog hepatocytes cultured on 200 collagen-coated borosillicate glass plates and found that albumin synthesis, ureogenesis, and gluconeogenesis occurred for 2 weeks. In addition, hemoperfusion of anhepatic dogs with this module improved the partial thromboplastin time (PTT), ammonia level, and amino acid levels, and also prolonged the survival time to 65 h. Dunn et al. [29] reported that rat hepatocytes cultured in a collagen sandwich configuration maintained a normal morphology and a physiological rate of albumin synthesis for at least 42 days. Clinically applicable liver support devices utilizing this culture system would have to be very large to contain a sufficient number of cultured hepatocytes, which is a major impediment to the development of ALS systems using monolayer cultured hepatocytes.

Immobilized hepatocytes in hydrogel

Lim et al. reported on the function of pancreatic islets encapsulated in calcium alginate gel [30]. Miura et al. evaluated the metabolic activity of isolated rat hepatocytes entrapped in this gel and demonstrated long-term maintenance of the synthesis of urea, albumin, and glucose, as well as the ability to detoxify hepatic toxins, e.g., phenols and fatty acids [31]. We transplanted encapsulated hepatocytes into the peritoneal cavity of rats with D-galactosamine-induced acute hepatic failure and found a marked improvement of liver function and survival [32].

Coating of entrapped hepatocytes with poly-L-lysine prolonged urea and albumin synthesis by a few days [33], and intraperitoneal transplantation of these cells into Gunn rats or rats with D-galactosamine-induced acute hepatic failure achieved a decrease in the bilirubin level and a prolongation of survival, respectively [34]. The major limitation of biological ALS systems incorporating isolated hepatocytes is the gradual loss of cellular function with time. Dixit et al. enveloped hepatocytes encapsulated in a matrigel with a sodium alginate-poly-L-lysine-sodium alginate copolymer membrane, which maintained hepatocellular function for an extended period, and transplanted the hepatocytes intraperitoneally into Gunn rats [35]. These transplanted, encapsulated hepatocytes were able to control the total serum bilirubin level for 28 days. In addition, Matthew et al. reported that in vitro perfusion of rabbit hepatocytes encapsulated in cellulose-chondroitin sulfate maintained ammonia, diazepam, and antipyrine metabolism for 10 days [36].

Iwata et al. attempted to develop an ALS system with agarose-gel-entrapped hepatocytes in the extracapillary space of a hollow fiber module (H. Iwata, Y. Ikada; personal communication). Takeshita et al. packed collagen-gel-entrapped hepatocytes into the extracapillary space of a hollow fiber module and reported that the hepatocytes synthesized twice as much albumin as monolayer-cultured hepatocytes and showed 43% of their ammonia removal on day 0 even after 8 days of culture [37]. Nyberg et al. developed a polysulfone hollow fiber cartridge with a nominal molecular weight cutoff of 100 kDa, and reported that glucose metabolism was maintained, blood amino acids were normalized, and lidocaine was metabolized during extracorporeal perfusion of anhepatic rabbits with the cartridge containing collagen-gel entrapped rat hepatocytes in its intracapillary spaces [38].

Matrices

It is well known that hepatocytes function better when attached to a matrix. Therefore, culture of cell lines and primary cells on dextran microcarriers (cytodex) was investigated by van Wezel [39], and long-term maintenance of cellular function and growth was achieved. Demetriou et al. reported that type I collagen coating of the microcarriers improved cell attachment and that intraperitoneal transplantation of hepatocytes attached to microcarriers decreased the serum bilirubin level in Gunn rats, as well as increasing the plasma albumin levels in Nagase analbuminemic rats (NAR) and improved the survival rate of rats with acute hepatic failure [40]. They then developed a hollow fiber bioreactor containing $4-6 \times 10^9$ microcarrier-attached porcine hepatocytes in the extracapillary spaces and 10 out of 10 patients with fulminant hepatic failure were successfully bridged to liver transplantation by plasma perfusion using this system (Fig. 5). However, only two out of six patients with CHF were relieved. This indicates that their system only seems to be effective for short-term bridging use. Shinyra et al. reported that the survival rate of rats with D-gal- or CCl_4-induced acute hepatic failure improved from 5% to 60% after direct hemoperfusion with newly developed polystyrene microcarrier-attached hepatocytes packed into a small column [41]. Yanagi et al. studied the metabolic activity of hepatocytes in a packed-bed reactor loaded with multiporous polyvinyl formal resin and found that ammonia metabolism and ureogenesis persisted for 2 weeks, and albumin secretion was observed for 20 days, significantly longer compared with that in monolayer culture [42]. We have found that the metabolic activity of rat hepatocytes attached to collagen-coated multiporous cellulose microcarriers can be preserved for at least 1 week [43], and we are currently attempting to develop a compact module for this hepatocyte system. Thus, these microcarriers are expected to achieve the long-term maintenance of hepatocyte function, and may contribute to the future development of compact modules for ALS.

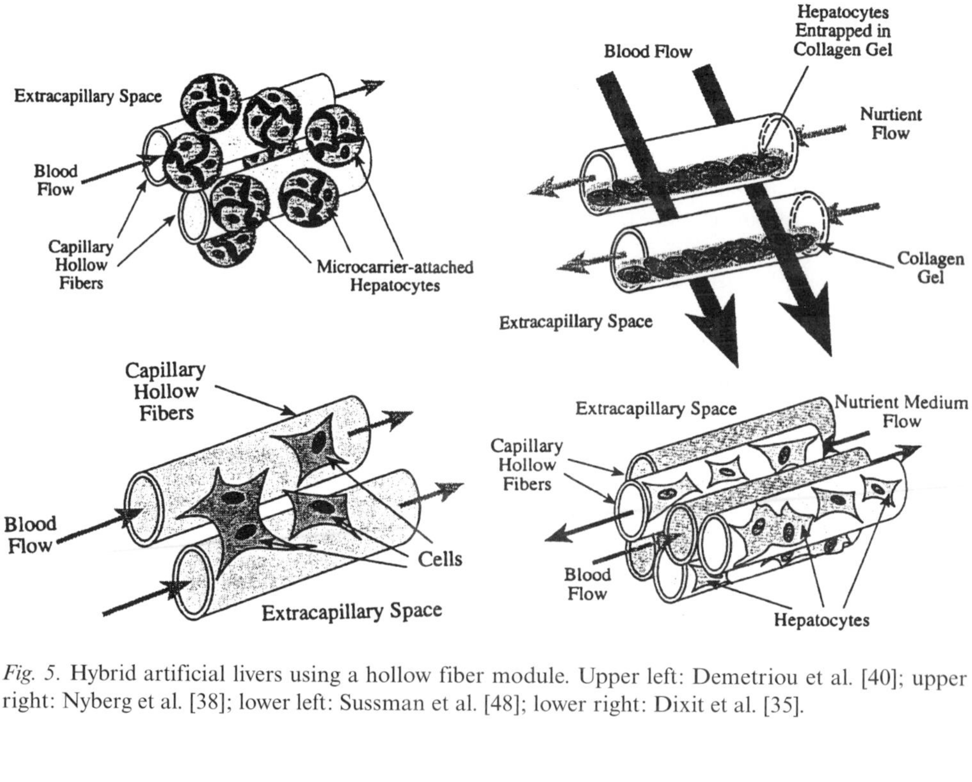

Fig. 5. Hybrid artificial livers using a hollow fiber module. Upper left: Demetriou et al. [40]; upper right: Nyberg et al. [38]; lower left: Sussman et al. [48]; lower right: Dixit et al. [35].

Hollow fiber modules

In hybrid ALS systems, hollow fiber membranes can have two different functions; one is their action as a membrane separator, and the other is for use as a hepatocyte culture system (Fig. 5). Knazek and Gullino cultured H4-11-EC3 hepatoma cells in the extracapillary space of artificial capillaries made of acrylic copolymer fibers with a molecular weight cutoff of 50,000 Da [44], and hemoperfusion of Gunn rats with this system achieved bilirubin conjugation [45]. Collagen or PVLA was used to coat hollow fiber membranes to allow hepatocyte attachment with long-term maintenance of metabolic activity [46,47]. Sussman et al. developed an extracorporeal liver assist device containing a highly differentiated human hepatoblastoma cell line (C3A cells) cultured on a cellulose acetate hollow fiber membrane, which supported dogs with acetaminophen-induced fulminant hepatic failure for long enough to allow liver regeneration [48]. A total of 11 patients have already been treated with this system. They reported no short-term safety problems. In addition, metabolic support was documented in 10 of the 11 patients, and six patients had a successful outcome. The C3A cells show a high proliferative capacity, but seem to have limited hepatocellular function. Recently, Gerlach et al. inserted the three to four woven capillary membrane systems in one cartridge to mimic the three-dimensional structure of the liver [49]. These capillary systems are used for culture medium supply, oxygen supply, nonparenchymal liver cells (Kupffer cells), hepatocytes, and blood circulation. This type of ALS system is capable of maintaining the metabolic activity of liver cells for a long period while avoiding immunological hazards and is also a compact device.

Hepatocyte spheroids

Monolayer cultures of hepatocytes provide a two-dimensional environment that may hamper some of their normal regulatory processes and functional activities. Hepatocytes exist in a three-dimensional environment in vivo, and cell-to-cell interactions can regulate the growth as well as the functional activity of isolated cells [50]. Therefore, the survival and functional activity of isolated hepatocytes may be improved by a three-dimensional environment, and many efforts have been made to produce three-dimensional aggregates of hepatocytes or hepatocyte spheroids.

Koide et al. reported that adult rat hepatocytes cultured in the presence of proteoglycans from multicellular spheroids showed a higher metabolic activity than that of cells in the usual culture system [51]. Following their experiments, many investigators have reported good results with systems using hepatocyte spheroids instead of two-dimensional systems. Koide et al. also encapsulated hepatocyte spheroids in calcium-alginate gel to avoid the formation of excessively large spheroids and to prevent cell detachment; the spheroids were then loaded into a culture chamber. During 48 h of perfusion, the cells maintained a normal morphology and showed high levels of albumin and urea synthesis [52]. Ito and Chang found that hepatocyte spheroids formed within 48 h when cells were encapsulated in alginate-poly-L-lysine, and these spheroids continued to secrete albumin throughout the 30 days of observation [34]. Hepatocyte spheroid formation was also achieved in PVLA-coated culture dishes [47], as well as in the pores of PVLA-coated reticulated polyurethane [53] or porous polyurethane foam (PUF) [54]. Hepatocyte spheroids immobilized in PUF showed the same rate of urea synthesis as cells in vivo, and spheroids in a culture column maintained urea synthesis for 1 month. Matsushita et al. evaluated the function of a newly developed PUF/spheroid packed-bed type ALS system in rats with hepatic failure [55]. The system decreased the blood ammonia level and four out of five rats survived. Further refinement of these systems is expected to assist the development of effective hybrid ALS systems.

Future prospects for hybrid artificial livers

Although many problems have already been solved due to developments in tissue culture and bioengineering techniques, there are still some other problems that need to be overcome before the hybrid ALS systems can be clinically applied. Immunological hazards can be avoided by encapsulation techniques, use of semipermeable hollow fiber membranes, and immunosuppressants. Long-term maintenance of the metabolic activity of hepatocytes can be achieved by immobilization in matrices or microcapsules. Recent advances in cryobiology have contributed to the long-term preservation of various biomaterials in large quantities [56], and the use of differentiated cell lines from human livers which has also enabled the preparation and preservation of cells in large numbers. Isolated hepatocytes or cells from differentiated liver cell lines can be cultured at a density of 10^7/ml, but for a clinically applicable ALS systems the cell density needs to be increased to 10^8/ml. Therefore, much effort is still needed to achieve the effective treatment of hepatic failure with ALS systems.

References

1. Malchesky PS. Nonbiological liver support: historic overview. Artif Organs 1994;18:342–347.
2. Bismuth H, Samuel D, Castaing D, Adam R, Saliba F, Johann M, Azoulay D, Ducot B, Chiche L. Orthotopic liver transplantation in fulminant and subfulminant hepatitis. The Paul Brousse experience. Ann Surg 1995;222:109–119.
3. Abel JJ, Rowntree LG, Turner BB. On the removal of diffusable substances from the circulating blood of living animals by dialysis. J Pharm Exp Ther 1914;5:275–316.
4. Killey JE, Pender JC, Welch HF, Welch CS. Ammonia intoxication treated by hemodialysis. N Engl J Med 1958;259:1156–1161.
5. Krebs HA, Henselleit K. Untersuchungen ueber die Harnstoff bildung im Tier Koerper. Z Physiol Chem 1932;210:33–66.
6. Kimoto S. The artificial liver, experiments and clinical application. Trans ASAIO 1959;5: 102–112.
7. Sorrentino F. Prime ricerche per la realizzazione di un fegato artificiale. Chir Patol Sper 1956;4:

1401–1414.
8. Mikami J, Mito M, Nishimura A, Sasaki E, Nose Y, Tedo I. Surgical treatment of acute liver failure. 2. An experimental study of extracorporeal metabolism in the artificial liver using slices of canine liver. Jpn J Gastroenterol 1959;56:1022.
9. Schechter DC, Nealon TF, Gibbon JH Jr. A simple extracorporeal device for reducing elevated blood ammonia levels. Surgery 1958;44:892–897.
10. Yatzidis H. A convenient hemoperfusion microapparatus over charcoal for the treatment of endogenous and exogenous intoxication. Proc Eur Dial Transpl Assoc 1964;1:83–874.
11. Gazzard BG, Weston MJ, Murray-Lyon IE. Experience at King's College Hospital with charcoal haemoperfusion; overall results in 37 patients. In: Williams R, Murray-Lyon IM (eds) Artificial Liver Support. London: Pittman Medical, 1975;234–241.
12. Trey C. The fulminant hepatic failure surveillance study: brief review of the effects of presumed etiology and age on survival. CMAJ 1972;106:525–526.
13. Geiger A, Yamazaki S. Cystidine and uridine requirements of the brain. J Neurochem 1956;1: 98–100.
14. Zimmerman HJ. Hepatic Failure. In: Gall EA, Mostofi FK (eds) The Liver. Baltimore: The Williams & Wilkins Co., 1973;384–405.
15. Zieve L. Synergism among toxic factors and other endogenous abnormalities in hepatic encephalopathy. In: Brunner G, Schmidt FW (eds) Artificial Liver Support. Berlin: Springer Verlag, 1981;18–24.
16. Hughes RD, Williams R. Development of a blood compatible resin for artificial liver support. In: Brunner G, Schmidt FW (eds) Artificial Liver Support. Berlin: Springer Verlag, 1981; 89–95.
17. Chang TMS. Hemoperfusion, exchange transfusion, cross circulation, liver perfusion, hormones and immobilized enzymes. In: Brunner G, Schmidt FW (eds) Artificial Liver Support. Berlin: Springer Verlag, 1981;126–133.
18. Inoue N, Horiuchi T, Hirose S, Funaki H, Yamazaki A, Yoshiba M, Wada T, Oda T, Imamiya T, Ide M. Plasmapheresis in acute hepatic failure. Jpn J Artif Org 1982;11:710–714.
19. Opolon P. Lange-pore hemodialysis in fulminant hepatic failure. In: Brunner G, Schmidt FW (eds) Artificial Liver Support. Berlin: Springer Verlag, 1981;141–146.
20. Gimson AES, Ede R, Braude S, Hughes RD, Langley PJ, Williams R. Fulminant hepatic failure and artificial liver support. Gastroenterol Jpn 1982;17:144–162.
21. Opolon P, Huguet C, Granger A, Gallot D, Bloch P, and Bidallier M. Comparison of single and cross haemodialysis with a donor through cuprophan and a new polymer membrane. In: Williams R, Murray-Lyon (eds) IM: Artificial liver support. London: Pitman Medical, 1975; 186–201.
22. Yoshiba M, Sekiyama K, Iwamura Y, Sugata F. Development of reliable artificial liver support (ALS) – plasma exchange in combination with hemodiafiltration using high-performance membranes. Dig Dis Sci 1993;38:469–476.
22. Otto JJ, Pender JC, Cleary JH, Sensenig DM, Welch CS. The use of a donor liver in experimental animals with elevated blood ammonia. Surgery 1958;43:301–309.
24. Ozawa K, kamiyama Y, Ukikusa M, Tobe T. Significance of hepatic energy charge and blood ketone body ratio as criteria of liver support. Indication and efficiency in hepatectomized patients during cross dialysis. In: Brunner G, Schmidt FW (eds) Artificial Liver Support. Berlin: Springer Verlag, 1981;164–169.
25. Matumura KN, Guevara GR, Huston H, Hamilton WL, Rikimaru M, Yamasaki G, Matumura MS. Hybrid bioartificial liver in hepatic failure; preliminary clinical report. Surgery 1987;101: 99–103.
26. Kasai S, Oikawa I, Asakawa H, Yamamoto T, Sawa M, Sakata H, Kondou K, Munakata T, Mito M, Tanzawa H. Evaluation of an artificial liver support device using isolated hepatocytes. In: Atsumi K, Maekawa M, Ota K (eds) Progress in Artificial Organs. Cleveland: ISAO Press, 1983;734–738.

27. Kasai S. What is needed for an artificial liver? In: Nose Y, Kjellstrand C, Ivanovich P (eds) Progress in Artificial Organs 1985. Cleveland: ISAO press, 1986;711–714.
28. Hase Y. A hybrid artificial liver system composed of primary cultured canine hepatocytes. Jpn J Surg 1990;91:214–222.
29. Dunn JCY, Yarmush ML, Koebe HG, Tompkins RG. Hepatocyte function and extracellular matrix geometry; long term culture in a sandwich configuration. FASEB 1989;3:174–177.
30. Lim F, Sun AM. Microencapsulated islets as bioartificial endocrine pancreas. Science 1980;210: 908–910.
31. Miura Y, Akimoto T, Fuke Y, Yamazaki S, Yagi K. In vitro maintenance of terminal-differentiated state in hepatocytes entrapped within calcium alginate. Artif Organs 1987;11:361–365.
32. Hirai S, Kasai S, Mito M. Encapsulated hepatocyte transplantation for the treatment of D-galactosamine-induced acute hepatic failure in rats. Eur Surg Res 1993;25:193–202.
33. Sun AM, O'Shea GM. Microencapsulation of living cells — a long-term delivery system. J Contr Rel 1985;2:137–141.
34. Ito Y, Chang TMS. In vitro study of multicellular hepatocyte spheroids formed in microcapsules. Artif Organs 1992;16:422–426.
35. Dixit V, Arthur M, Reinhardt R, Gitnick G. Improved function of microencapsulated hepatocytes in a hybrid/bioartificial liver support system. Artif Organs 1992;16:336–341.
36. Matthew HWT, Salley SO, Peterson WD, Deshmukh DR, Mukhopadhyay A, Klein MD. Microencapsulated hepatocytes — prospects for extracorporeal liver support. Trans ASAIO 1991;37: 328–330.
37. Takeshita K, Ishibashi H, Suzuki M, Yamamoto T. Akaike T, Odama M. High cell-density culture system of hepatocytes entrapped in a three-dimensional hollow fiber with collagen gel. Artif Organs 1995;19:191–193.
38. Nyberg SL, Shirabe K, Peshwa MV, Sielaff TD, Crotty PL, Mann HJ, Remmel RP, Payne WD, Hu WS and Cerra FB. Extracorporeal application of a gel-entrapment, bioartificial liver-demonstration of drug metabolism and other biochemical functions. Cell Transpl 1993;2: 441–452.
39. van Wezel AL. Growth of cell-strains and primary cells on micro-carriers in homogeneous culture. Nature 1967;216:64–65.
40. Demetriou AA, Whiting J, Levenson SM, Chowdhury NR, Schechner R, Michalski S, Feldman D, Chowdhury JR. New method of hepatocyte transplantation and extracorporeal liver support. Ann Surg 1986;204:259–271.
41. Shinyra A, Bocharov A, Bochkova N, Spirov V. Bioartificial liver using hepatocytes on Biosilon microcarriers; treatment of chemically induced acute hepatic failure in rats. Artif Organs 1991;15:189–197.
42. Yanagi K, Miyoshi H, Fukuda H, Ohshima N. A packed-bed reactor utilizing porous resin enables high density culture of hepatocytes. Appl Microbiol Biotechnol 1992;37:316–320.
43. Kasai S, Sawa M, Nishida Y, Onodera K, Hirai S, Yamamoto T, Mito M. Cellulose microcarrier for high-density culture of hepatocytes. Transpl Proc 1992;24:2960–2961.
44. Knazek RA, Gullino PM. Dual circuit, woven artificial capillary bundle for cell culture. US Patent 4,184,922, 22 Jan, 1980.
45. Wolf CFW. Liver assist by cultured cells on artificial capillaries. In: Brunner G, Schmidt FW (eds) Artificial Liver Support. Berlin: Springer-Verlag, 1981;247–253.
46. Jauregui HO, Muller TE. Long-term cultures of adult mammalian hepatocytes in hollow fibers as the cellular component of extracorporeal (hybrid) liver assist devices. Artif Organs 1992;16: 209–212.
47. Tobe S, Takei Y, Maeda A, Akaike T, Kobayashi K. Formation of multilayer aggregation of adult rat hepatocytes in primary culture on substratum as asialoglycoprotein model. Jpn J Art Org 1990;19:1156–1160.
48. Sussman NL, Chong MG, Koussayer T, He DE, Shang TA, Whisennand HH, Kelly JH. Reversal of fulminant hepatic failure using an extracorporeal liver assist device. Hepatology 1992;16:

60–65.
49. Gerlach JC, Schnoy N, Encke J, Smith MD, Müller E, Neuhaus P. Improved hepatocyte in vitro maintenance in a culture model with woven multicompartment capillary systems – electron microscopy studies. Hepatology 1995;22:546–552.
50. Landry J, Freyer JP. Regulatory response in spheroidal aggregates of normal and cancerous cells. Recent Results Cancer Res 1986;95:50–66.
51. Koide N, Shinji Y, Tanabe T, Asano K, Kawaguchi M, Sakaguchi K, Koibe Y, Mori M, Tsuji T. Continued high albumin production by multicellular spheroids of adult rat hepatocytes formed in the presence of liver-derived proteoglycans. Biochem Biophys Res Commun 1989;161: 385–391.
52. Takabatake H, Koide N, Sakaki S, Matsushima M, Takenami T, Ono R, Sakaguchi K, Funatsu K, Tsuji T. Bioartificial liver using encapsulated hepatocyte spheroids. Jpn J Art Org 1991;20: 139–144.
53. Sato Y, Ochiya T, Matsubara K. A new three-dimensional culture system of rat hepatocytes using reticulated polyurethane. Jpn J Art Org 1991;20:145–149.
54. Ijima H, Matsushita T, Funatsu K. Development of a hybrid type artificial liver using PUF/ spheroid culture system of adult hepatocytes. Jpn Art Org 1993;22:171–176.
55. Matsushita T, Ijima H, Wada S, Funatsu K. Estimation of the performance of PUF Spheroid packed-bed type artificial liver by using an extracorporeal circulation with hepatic failure rats. Jpn J Art Org 1995;24:815–820.
56. Kasai S, Mito M. Large-scale cryopreservation of isolated dog hepatocytes. Cryobiology 1993; 30:1–11.

Tissue Engineering for Therapeutic Use 1.
Y. Ikada and Y. Yamaoka, editors.

Clinical application of bioartificial pancreas

Takashi Maki

Department of Surgery, Beth Israel Deaconess Medical Center and Harvard Medical School, Boston, Massachusetts, USA

Keywords: bioartificial pancreas, clinical trial, diffusion devices, immune exclusion, insulin treatment, microcapsules, pancreas transplantation, type 1 diabetes, vascularized devices.

Introduction

Patients with insulin-dependent diabetes mellitus (IDDM) continue to develop serious vascular, neurological and nephrological complications when they are treated by conventional insulin therapy (one or two insulin injections per day) combined with diet and exercise. The recent Diabetes Control and Complication Trial (DCCT) has unequivocally shown the importance of maintaining blood glucose levels close to the normal range in delaying the onset and slowing the progression of diabetic complications [1]. The insulin therapy in DCCT requires three or more insulin injections daily (or use of an insulin pump) with frequent dosage adjustments based on at least four finger-stick blood glucose monitoring. A major setback of the intensive insulin therapy is a 2- to 3-fold higher incidence of severe hypoglycemia than that observed in patients treated with conventional therapy. Moreover, primary-care physicians who treat the majority of diabetic patients may not be able to offer as rigorous and comprehensive treatment of diabetes as provided by a team of well-trained diabetes professionals in the DCCT [2].

With the emergence of effective immunosuppressive drugs and improved surgical techniques, pancreas transplantation has become a practical therapy for IDDM. Following successful transplantation, patients become insulin independent with full restoration of normal glucose metabolism. However, the necessity for chronic immunosuppression subjects patients to significant morbidity and mortality due to opportunistic infections and spontaneous neoplasms as well as direct drug toxicity and metabolic complications. The necessity for chronic immunosuppression would essentially replace one chronic illness (diabetes) with another (immunosuppression). Thus, many believe that only patients with severe diabetic complications can be justifiably transplanted with the pancreas under

Address for correspondence: Takashi Maki MD PhD, Department of Surgery, Beth Israel Deaconess Medical Center and Harvard Medical School, Research North/BIDMC, P.O. Box 15707, Boston, MA 02215, USA. Tel.: +1-617-632-0276. Fax: +1-617-632-0271.
E-mail: tmaki@west.bidmc.harvard.edu

immunosuppression. Ironically, the pancreas transplanted alone in nonuremic diabetic patients has a greater tendency to be rejected in spite of high dose immunosuppression than the pancreas simultaneously transplanted with kidney in diabetic patients with chronic renal failure [3]. The shortage of human donor pancreases also limits application of this treatment modality. Clinical islet transplantation has been performed only with limited success and also suffers from the chronic use of immunosuppression and a shortage of pancreas donors.

The use of bioartificial pancreases capable of excluding immune cells and immunoglobulins, thus eliminating the necessity for chronic immunosuppression, has emerged as a potential therapeutic modality for IDDM. In all three types of bioartificial pancreas devices, i.e., microcapsules, diffusion devices, and perfusion-based vascular devices, the immune exclusion is achieved by separating islets from the host immune system by semipermeable membranes that allow only molecules of a defined size to pass through. Small molecules such as glucose, insulin, nutrients and metabolites are exchanged across the membrane, whereas lymphocytes and immunoglobulins are excluded by the membrane and thus unable to cause rejection of the islets. Thus, the bioartificial pancreas offers the advantages of islet transplantation without immunosuppression and the possibility of circumventing the shortage of human islets by using xenogeneic islets. Furthermore, since prevention of the diabetes complications probably requires early intervention to correct abnormal glucose metabolism, the bioartificial pancreas which does not require immunosuppression may be ideal for early treatment of IDDM.

In this article I will briefly review the preclinical (large animal) and clinical application of microcapsules and diffusion devices by other investigators. I will then summarize our group's experience in the use of a vascularized bioartificial pancreas for treatment of experimental diabetes in dogs and discuss plans to initiate a clinical trial of the vascularized bioartificial pancreas for treatment of patients with early stage IDDM.

Microencapsulation

The major interest in islet transplantation by immune exclusion methods continues to be microencapsulation. With microencapsulation, a single islet or small numbers of islets are encapsulated within gel membranes and implanted into the peritoneal cavity or some other anatomic sites. Insulin secreted by the encapsulated islets diffuses across the membrane to the extracellular fluid and is eventually absorbed into the circulation. Inflammation around the microcapsules with tissue adherence and pericapsular fibrosis continues to be a significant problem associated with microencapsulation techniques. Recent studies in small animals have focused on the improvement of materials and stability of the capsular membrane. These include analyses of foreign body reaction to the capsules [4—7], production of mechanically stable microcapsules by using alginate high in a guluronic acid content [8,9], use of hydrophilic macrobeads made with various combinations of agarose, collagen, and gel-foam [10], use of microbeads composed of an agarose and polystyrene sulfonic acid mixed gel [11], and an attempt to use uncoated calcium alginate microspheres [12,13]. Coherent microcapsules which tightly envelop each individual islet have been utilized in an attempt to reduce the overall volume for implantation [14].

There are a limited number of large animal studies using microencapsulation techniques. Soon-Shiong and colleagues [15,16] implanted allogeneic islets (15,000–20,000 islets/kg) encapsulated within low mannuronic acid/high guluronic acid alginate microcapsules intraperitoneally into spontaneously diabetic dogs under cyclosporine (CsA) immunosuppression (10 mg/kg/day). All of the animals (n = 6) became normoglycemic and remained free of exogenous insulin for a period of 63–172 days with significant improvements in serum glycosylated hemoglobin (HbA1c) levels, intravenous glucose tolerance (K-values) and C-peptide release. The animals (n = 3) which received second encapsulated islet implantation following recurrence of hyperglycemia restored normoglycemia for an additional 76–130 days. Overall graft survival as defined by the presence of circulating C-peptide had reached 726 days. Wahoff et al. [17] showed that canine islets encapsulated within the same type of microcapsules maintained glycemic control for up to 156 days in pancreatectomized diabetic dogs under low-dose CsA immunosuppression. Sun et al. [18] reported normalization of spontaneous diabetes in cynomolgus monkeys by implantation of microencapsulated porcine islets (20,000–70,000 islets/animal) without immunosuppression. Insulin independence with restoration of normoglycemia, improved glucose tolerance, circulating porcine C-peptide and decreased HbA1c values were observed for 106–390 days. In an attempt to provide a higher oxygen tension to encapsulated islets, Calafiore et al. [19,20] placed xenogeneic (human) islets encapsulated in sodium alginate microcapsules between the two layers of coaxial vascular prostheses (inner permeable Dacron® and outer impermeable PTFE®) which was then anastomosed to iliac vessels as an arteriovenous (A-V) shunt in alloxan-induced diabetic dogs. No immunosuppression was used. Following implantation (40,000 islets/animal), recipient dogs showed a long-term decline in exogenous insulin requirement along with near normalization of blood glucose levels and significant circulating human C-peptides.

Calafiore [20] used microcapsules embodied in a vascular coaxial prosthesis in two patients; a type II diabetic requiring an axillary-femoral arterial bypass and a type I diabetic who had rejected a whole pancreas transplant. In the type II diabetic patient, implantation of 150,000 human islets achieved a 5-fold decrease in exogenous insulin requirement and an increase in baseline and stimulated C-peptide release. Islet function deteriorated after 8 months. In the second patient, implantation of 250,000 human islets with a creation of an axillary A-V bypass led to the appearance of serum C-peptide associated with improvement of blood glucose control and a small reduction of exogenous insulin requirement during 60 days of clinical follow up. Soon-Shiong et al. [21] subsequently reported a clinical case in which encapsulated allogeneic islets were used to treat a type I diabetic who was on low-dose CsA and azathioprine immunosuppression after a successful living-related kidney transplant. In total, two intraperitoneal injections of microcapsules (9,957 and 5,000 islets) given 6 months apart achieved tight glycemic control with much reduced exogenous insulin requirement for up to 9 months. In the 9th month, exogenous insulin could be discontinued with stable blood glucose levels. No reports have been available concerning the follow up of this case or additional clinical trials.

Diffusion chambers

In diffusion chambers, a large number of islets are enclosed in semipermeable membranes of various geometric configurations with various volumes, e.g., straw-shape tubular membranes [22,23], acrylic copolymer hollow fibers [24,25], hockey-puck-shape bimembrane discs [26], mesh reinforced polyvinyl alcohol hydrogel bags [27], etc. The islets are suspended in a hydrogel matrix such as agarose, alginate and collagen gel. Diffusion chambers are usually placed in intraperitoneal or subcutaneous sites. Similar to microcapsulation, diffusion devices depend on nutrients and oxygen which diffuse from the interstitial fluid.

There is only one large animal study with diffusion chambers. We, together with Lanza's group, used slightly larger tubular diffusion chambers (straws) than those employed in rats [22] in totally pancreatectomized diabetic dogs without immunosuppression [28]. Following implantation of up to 200 straws containing canine islet allografts (n = 6), excellent long-term glucose control with significant improvement in IVGTT was achieved with complete elimination of exogenous insulin for 46–86 days in three dogs. Eventually glucose control was lost. Laparotomy revealed extensive breakage of straws but relatively minimal tissue reaction. While these studies illustrated the potential usefulness of diffusion chambers, the limitations imposed by fragile structures and requirement for a large volume were also evident.

Scharp and colleagues [29] used permselective acrylic-copolymer hollow fibers (1.5 cm $\times \sim$800 mm) to encapsulate human islet allografts (50 IE/fiber) and transplanted them (3 seeded and 1 unseeded fiber) subcutaneously in three groups of patients: type I diabetics of $>$20 year duration (n = 3), type II diabetics of $>$5 year duration (n = 3) and human nondiabetic volunteers (n = 3). After 2 weeks in situ, the fibers were removed and examined for insulin secretion and viability. While there was no effect on glucose homeostasis, the fibers produced minimal or no discomfort in the patients. Glucose responsive insulin release was present in encapsulated islets removed from normal control individuals but was absent in islets removed from type I and type II diabetic patients. All groups of encapsulated islets were responsive to glucose plus theophylline stimulation and maintained approximately 90% viability. These short-term preliminary, preclinical studies demonstrated that macroencapsulated human islet allografts can survive in subcutaneous locations protected from allogeneic immune and autoimmune responses.

Vascularized bioartificial pancreas

Our group has been interested in the use of a perfusion-based vascularized bioartificial pancreas for treatment of diabetes in experimental animals [30–34]. The device consists of a disk-shaped, acrylic housing and an internal single-coiled acrylic copolymer membrane tube. The membrane has a nominal molecular weight cutoff of approximately 70 kDa. The islets are usually suspended in a hydrogel matrix such as agar or alginate to prevent settling and to assure even distribution, and seeded into an annular space between the tubular membrane through a seeding port. Both ends of the tubular membrane are connected to PTFE vascular grafts with an internal diameter of 6 mm which are anastomosed to iliac vessels of the host animals as an A-V shunt. Thus, the vascularized bioartificial pancreas has the advantage of a high oxygen tension of the arterial blood exposed to the islets. When they were implanted in normal dogs, the devices were patent for a period ranging from 8 months to more than 3.5 years [35], illustrating their excellent biocompatibility.

Our group was the first to demonstrate the feasibility of a bioartificial pancreas in large animals [30–33]. Using a severe brittle form of diabetes induced by total pancreatectomy in the dog [36], we showed that glycemic control for up to 1 year with zero or minimal exogenous insulin could be achieved by implanting in a recipient dog two vascularized bioartificial pancreas devices containing allogeneic (canine) islets. At no time was immunosuppression required, thus confirming the immune-exclusive nature of the devices. One of these devices that was removed after a period of 1 year showed glucose-stimulated insulin release upon reperfusion in vitro.

More recently, we demonstrated that a single vascularized artificial pancreas containing porcine islet xenografts (160,000–430,000 IE/device) achieved long-term glycemic control in a pancreatectomized dog [34]. Again no immunosuppression was used. Following implantation 7 out of 17 dogs maintained fasting blood glucose levels at <200 mg/dl for 53–263 days with 4–16 units/day of exogenous insulin compared to a preimplant insulin requirement of 31–44 units/day (a 54–74% overall reduction). Varying levels of porcine C-peptide were detected during the same period. Of the dogs, two maintained good glycemic control for more than 8 months with improved glucose tolerance and a significant time-dependent reduction in HbA1c levels. Although these dogs continued to have a high titer of naturally occurring antiporcine cytotoxic antibody, survival of porcine islets contained in the devices was not affected. These results provide unequivocal evidence that the vascularized bioartificial pancreas containing xenogeneic porcine islets is a promising therapeutic modality for treatment of IDDM.

Plans for clinical phase I/II trial of vascularized bioartificial pancreas containing porcine islets

Our results have clearly demonstrated that a vascularized bioartificial pancreas overcomes two major problems associated with pancreatic transplantation, i.e., complications associated with the chronic use of immunosuppression and the limited availability of human donor organs. Moreover, since immunosuppression is not required, the vascularized bioartificial pancreas may be justifiably used to treat patients before they develop severe diabetic complications. Encouraged by our large animal studies, we are currently planning a clinical phase I/II trial of the vascularized artificial pancreas. This will be a single site, open-label study of safety, tolerability and potential efficacy of a bioartificial pancreas seeded with porcine islets to be conducted in six patients. Patients included in the study will be labile type 1 diabetics who suffer from frequent debilitating hypoglycemic episodes as defined by the DCCT Research Group [37] despite supervised medical treatment, and those who have not developed serious vascular, neurological and nephrological complications. Patients with renal failure, any organ transplant, known insulin resistance, or evidence of significant cardiac or peripheral vascular disease will be excluded from the study. The patients who continue to have two or more episodes of severe hypoglycemia within 2 months of the screening period in spite of optimization of medical management of their diabetes will be offered the opportunity for implantation of the bioartificial pancreas seeded with porcine islets. The device will be implanted at the iliac fossa in a similar fashion as renal transplantation. No immunosuppression will be used at any point in time. The patients will be given 325 mg/day aspirin for the prevention of clotting. Following implantation, the patients' cardiovascular and pulmonary functions, as well as the vascular patency of the device, will be monitored. Although this will be a phase I/II study, the patients' glucose and insulin requirements will also be monitored. Reduction of exogenous insulin requirements, less frequent hypoglycemic episodes, and/or the presence of porcine C-peptide will be considered as presumptive evidence of device function. The patients will be followed for as long as the device continues to function. The device will be removed if thrombosis or any other adverse event occurs.

Concluding remarks

As we shift our focus from the animal studies to the clinical application in an attempt to use a bioartificial pancreas for treatment of diabetes, it is important to recognize several potential limitations associated with the three types of the bioartificial pancreas devices.

While microcapsules have the advantage of easy handling with minor surgery for implantation, many important questions such as long-term membrane stability, the long-term effect of large amounts of the membrane material in the abdominal cavity, and the retrievability of microcapsules need to be answered. It is apparent from the literature that the oxygen tension of the peritoneal cavity, even though it is low, is sufficient to sustain long-term viability of encapsulated islets. The necessity or role of immunosuppression for survival of microencapsulated islets must also be determined.

Diffusion devices have the ease of minor surgical placement and retrieval especially when they are implanted subcutaneously. However, the use of the tubular and hollow fiber devices could be limited by the large volume/number required to achieve optimal glycemic control as well as long-term stability. Similarly, the obvious question of volume/number required for the disc and bag-type devices needs to be addressed in a large animal model prior to their clinical application.

The major limitation for the use of vascularized devices is the necessity for vascular surgery and subsequent vascular complications. Although our dog studies, particularly with the use of porcine islets, were complicated by vascular thrombosis, this problem may be unique to the canine species reflecting the well recognized difficulty in maintaining the patency of vascular grafts in dogs due to their blood hypercoagulability [38]. Patients may suffer from a peripheral "steal" syndrome caused by diversion of blood from the distal extremity through a large arteriovenous shunt in the device. Because of the larger size of the patients compared to the size of dogs, a higher insulin secretory capacity of the device may be necessary to achieve adequate glycemic control. Further studies are needed to maximize the insulin secretory capacity of the device and to develop a new islet suspension matrix which would enable periodic islet replacement in the device.

In conclusion, the immune exclusive nature combined with the use of xenogeneic islets makes the bioartificial pancreas a potentially superior therapeutic modality for treatment of diabetes when compared to other means of diabetes management. While complete elimination of exogenous insulin usage is desirable after implantation of a bioartificial pancreas, tight glycemic control achieved by small amounts of insulin given in a low number of injections would make the treatment of diabetes much less rigorous compared to DCCT-type intensive insulin treatment in preventing the development of diabetic complications. Continuing research is warranted for the development of a safer and more efficacious bioartificial pancreas.

References

1. The Diabetes Control and Complications Trial Research Group. The effect of intensive treatment of diabetes on the development and progression of long-term complications in insulin-dependent diabetes mellitus. N Engl J Med 1993;329:977–986.
2. Harris MI, Eastman RC, Siebert C. The DCCT and medical care for diabetes in the US. Diabet Care 1994;17:761–764.
3. Sutherland DER, Gruessner A, Moundry-Munns K. Analysis of united network for organ sharing (UNOS) United States of America (USA) pancreas transplant registry data according to multiple variables. In: Terasaki P, Cecka JM (eds) Clinical Transplants. California: UCLA Tissue Typing Laboratory, 1992;45–59.
4. Fritschy WM, DeVos P, Groen H, Klatter FA, Pasma A, Wolters GH, van Schilfgaarde R. The capsular overgrowth on microencapsulated pancreatic islet grafts in streptozotocin and autoimmune diabetic rats. Transpl Int 1994;7:264–271.
5. Cole DR, Waterfall M, McIntyre M, Baird JD. Microcapsulated islet grafts in the BB/E rat: a possible role for cytokines in graft failure. Diabetologica 1992;35:231–237.
6. Weber CJ, Zabinski S, Koschitzky T, Wicker L, Rajotte R, D'Agati V, Peterson L, Norton J,

Reemtsma K. The role of $CD4^+$ helper T cells in the destruction of microencapsulated islet xenografts in NOD mice. Transplantation 1990;49:396–404.

7. Weber CJ. Microencapsulated Islet Iso-, Allo-, and Xenografts in NOD and NOD-Scid Mice: Metabolic, Immunologic, and Bioengineering Studies. Presented at 7th Taniguchi Conference on Polymer Chemistry, Tissue Engineering with the Use of Biomedical Polymers, Kyoto, 1995.
8. Soon-Shiong P, Otterlei M, Skjak-Brack G, Smidsrod O, Heintz R, Lanza RP, Espevik T. Immunologic basis for the fibrotic reaction to implanted microcapsules. Transpl Proc 1991;23: 758–759.
9. Otterlei M, Ostgaard K, Skjak-Brack G, Smidsrod O, Soon-Shiong P, Espevik T. Induction of cytokine production from human monocytes stimulated with alginate. J Immunother 1991;10: 286–291.
10. Jain K, Yang H, Cai B-R, Hague B, Hurvitz AI, Diehl C, Miyata T, Smith BH, Stenzel K, Suthanthiran M, Rubin AL. Retrievable, replaceable, macroencapsulated pancreatic islet xenografts. Long-term engraftment without immunosuppression. Transplantation 1995;59:319–324.
11. Iwata H, Takagi T, Kobayashi K, Oka T, Tsuji T, Ito F. Strategy for developing microbeads applicable to islet xenotransplanation into spontaneous diabetic NOD mice. J Biomed Mater Res 1994;28:1201–1207.
12. Lanza RP, Kuhtreiber WM, Ecker D, Staruk JE, Chick WL. Xenotransplantation of porcine and bovine islets without immunosuppression using uncoated alginate microspheres. Transplantation 1995;59:1377–1384.
13. Lanza RP, Kuhtreiber WM, Ecker D, Staruk JE, Chick WL. Successful Bovine Islet Xenografts in Rodents and Dogs Using Injectable Microreactors. Fifth International Congress on Pancreas and Islet Transplantation. Miami Beach, Florida, 1995;103 (Abstract).
14. Calafiore R, Basta G, Osticioli L, Luca G, Tortoioli C, Brunetti P. Coherent microcapsules for pancreatic islet transplantation: a new approach for bioartificial pancreas. Transplant Proc 1996;28:812–813.
15. Soon-Shiong P, Feldman E, Nelson R, Komtebedde J, Smidsrod O, Skjak-Braek G, Espevik T, Heintz R, Lee M. Successful reversal of spontaneous diabetes in dogs by intraperitoneal microencapsulated islets. Transplantation 1992;54:769–774.
16. Soon-Shiong P, Feldman E, Nelson R, Heintz R, Yao Q, Yao Z, Zheng T, merideth N, Skjak-Braek G, Espevik T. Long-term reversal of diabetes by the injection of immunoprotected islets. Proc Natl Acad Sci USA 1993;90:5843–5847.
17. Wahoff DC, Stephanian E, Gores PF, Soon-Shiong P, Hower C, Lioveras JK, Sutherland DER. Intraperitoneal transplantation of microencapsulated canine islet allografts with short-term, low-dose cyclosporine for treatment of pancreatectomy-induced diabetes in dogs. Transplant Proc 1994;26:804.
18. Zhou D, Sun YL, Vacek I, Ma P, Sun AM. Normalization of diabetes in cynomolgus monkeys by xenotransplantation of microencapsulated porcine islets. Transplant Proc 1994;26:1091.
19. Calafiore R, Basta G, Falomi A Jr, Ciabattoni P, Brotzu G, Cortesini R, Brunetti P. Intravascular transplantation of microencapsulated islets in diabetic dogs. Transplant Proc 1992;24: 935–936.
20. Calafiore R. Transplantation of microencapsulated pancreatic human islets for therapy of diabetes mellitus: a preliminary report. ASAIO J 1992;38:34–37.
21. Soon-Shiong P, Heintz RE, Merideth N, Yao QX, Yao Z, Zheng T, Murphy M, Moloney MK, Scmehl M, Harris M, Medez R, Mendez R, Sandford P. Insulin independence in a type 1 diabetic patient after encapsulated islet transplantation. Lancet 1994;343:950–951.
22. Lanza RP, Butler DH, Borland KM, Staruk JE, Faustman DL, Solomon BA, Muller TE, Rupp RG, Maki T, Monaco AP, Chick WL. Xenotransplantation of canine, bovine and porcine islets in diabetic rats without immunosuppression. Proc Natl Acad Sci USA 1991;88:11100–11104.
23. Lanza RP, Borland KM, Staruk JE, Appel MC, Solomon BA, Chick WL. Transplantation of encapsulated canine islets into spontaneously diabetic BB/Wor rats without immunosuppression. Endocrinology 1992;131:637–642.

24. Lacy PE, Hegre OD, Gerasimidi-Vazeou A, Gentile FT, Dionne KE. Maintenance of normoglycemia in diabetic mice by subcutaneous xenografts of encapsulated islets. Science 1991;254: 1782–1784.
25. Ward RS, White KA, Wolcott CA, Wang AY, Kuhn RW, Taylor JE, John JK. Development of a hybrid artificial pancreas with a dense polyurethane membrane. ASAIO J 1993;39:M261–267.
26. Ohgawara H, Miyazaki J, Karibe S, Tashiro F, Akaike T, Hashimoto Y. Embedded-culture of pancreatic beta cells derived from transgenic mouse insulinoma as a potential source for xenotransplantation using a diffusion chamber. Cell Transplant 1995;4:307–313.
27. Hayashi H, Inoue K, Tun T, Aung T, Gu YJ, Kato M, Echitgo Y, Kaji H, Imamura M, Iwata H, Ikada Y, Miyazaki J. Xenotransplantation of β-Cell Line Min6 in Mesh Reinforced Polyvinyl Alcohol Hydrogel Bag. Fifth International Congress on Pancreas and Islet Transplantation. Miami Beach, Florida, 1995;101 (Abstract).
28. Lanza RP, Borland KM, Sullivan SJ, Lodge P, Carretta M, Muller TE, Solomon BA, Maki T, Monaco AP, Chick WL. Treatment of severely diabetic, pancreatectomized dogs using a diffusion-based hybrid pancreas. Diabetes 1992;41:886–889.
29. Scharp DW, Swanson CJ, Olack BJ, Latta PP, Hegre OD, Doherty EJ, Gentile FT, Flavin KS, Ansara MF, Lacy PE. Protection of encapsulated human islets implanted without immunosuppression in patients with type I or type II diabetes and in nondiabetic control subjects. Diabetes 1994;43:1167–1170.
30. Maki T, Ubhi CS, Sanchez-Farpon H, Sullivan SJ, Borland K, Muller TE, Solomon BA, Chick WL, Monaco AP. Successful treatment of diabetes with the biohybrid artificial pancreas in dogs. Transplantation 1991;51:43–51.
31. Sullivan SJ, Maki T, Borland KM, Mahoney MD, Solomon BA, Muller TE, Monaco AP, Chick WL. The biohybrid artificial pancreas: long-term implantation studies in diabetic, pancreatectomized dogs. Science 1991;252:718–721.
32. Monaco AP, Maki T, Ozato H, Carretta M, Sullivan SJ, Borland KM, Mahoney MD, Chick WL, Muller TE, Wolfrum J, Solomon B. Transplantation of islet allografts and xenografts in totally pancreatectomized diabetic dogs utilizing the hybrid artificial pancreas. Ann Surg 1991; 214:339–362.
33. Maki T, Lodge JPA, Carretta M, Ohzato H, Borland KM, Sullivan SJ, Staruk J, Solomon BA, Chick WI, Monaco AP. Treatment of severe diabetes mellitus for more than 1 year using a vascularized hybrid artificial pancreas. Transplantation 1993;55:713–718.
34. Maki T, Otsu I, O'Neil JJ, Dunleavy K, Mullon CJP, Solomon BA, Monaco AP. Treatment of diabetes by xenogeneic islets without immunosuppression: use of a vascularized bioartificial pancreas. Diabetes 1996;45:342–347.
35. Maki T, Mullon CJP, Solomon BA, Monaco AP. Novel delivery of pancreatic islet cells to treat insulin-dependent diabetes mellitus. Clin Pharmacokin 1995;28:471–482.
36. Maki T, Monaco AP, Mullon CJP, Solomon BA. Early treatment of diabetes with porcine islets in a bioartificial pancreas. Tis Eng 1996;2:299–306.
37. The Diabetes Control and Complications Trial Research Group. Epidemiology of severe hypoglycemia in the diabetes control and complications trial. Am J Med 1991;90:450–453.
38. Dewanjee MK, Kapadvanjwala M, Sanchez A, Elson R, Serafini AN, Zilleruelo GE, Sfankianakis GN. Quantization of comparative thrombogenicity of dog, pig, and human platelets in a hemodialyzer. ASAIO J 1992;38:88–90.

Tissue Engineering for Therapeutic Use 1.
Y. Ikada and Y. Yamaoka, editors.

Tissue engineering applied to cancer plastic and reconstructive surgery

Charles W. Patrick Jr and Gregory R.D. Evans
Department of Plastic Surgery, The University of Texas, M.D. Anderson Cancer Center, Houston, Texas, USA

Keywords: adipose, bone, neovascularization, peripheral nerve, plastic surgery, reconstructive surgery, tissue engineering.

Introduction

Tissue engineering has been heralded as one of the most exciting advances that is occurring in the field of plastic surgery [1]. In fact, at the inaugural meeting of the Tissue Engineering Society, Kiki Hellman listed the development of novel reconstructive surgery applications as one of the future challenges for tissue engineering [2]. The art of plastic surgery is the restoration of form and function with the resultant improvement in patient quality of life and aesthetic outcome. "We restore and make whole those parts which nature or ill fortune have taken away, not so much to delight the eye, but to buoy up the spirit of the afflicted", Gaspare Tagliacozzi [3]. Plastic is derived from the Greek word plastikos which means to mold or shape. While the term plastique was used by Desault in 1798, it was the publication of Zeis' "Handbuch der plastischen Chirurgie", in 1838 that popularized the term [4,5]. Many of the first plastic surgeries were developed to close difficult wounds or replace tissue loss due to injury or cancer. These procedures often involved the formation of flaps (a section of living tissue that carries its own blood supply and is moved from one area of the body to another to repair a deficit of skin, fat, muscle movement, or skeletal support) to reshape or mold the defect so as to approximate the original shape. In 1987 the National Science Foundation, USA, coined the term "tissue engineering", which was later defined as "...an interdisciplinary field that applies the principles of engineering and the life sciences toward the development of biological substitutes that restore, maintain, or improve tissue function" [6]. In many respects the terms tissue engineering and plastic surgery are synonymous. Both fields search for the restoration, maintenance, and improvement of tissue structure and function. "Not for self, not for the fulfillment of any earthly desire or gain but solely for the salvage of suffering humanity", Sushruta Samhita [7]. Today, 3,000 years later, the same

Address for correspondence: Dr Gregory R.D. Evans, Department of Plastic Surgery, The University of Texas, M.D. Anderson Cancer Center, 1515 Holcombe Blvd., Box 62, Houston, TX 77030, USA. Tel.: +1-713-794-1247. Fax: +1-713-794-5492. E-mail: Gregory.Evans@surg.mdacc.tmc.edu philosophy holds for both tissue engineering and plastic surgery. Both are dedicated to the reconstruction of tissue defects due to disease, trauma, or congenital abnormalities. Moreover, both fields are multidisciplinary in nature, with tissue engineering coupling bioengineering, life sciences, and surgical science, and plastic surgery coupling together the multiple fields of medicine.

From a historical standpoint, tissue engineering is the natural progression for the field of plastic surgery. In its infancy, plastic surgery transferred local and pedicle flaps to tissue defects. The introduction of the arm tubed pedicle flap was novel for its day (Fig. 1), providing the first evidence of tissue engineering applied to plastic surgery. An increased knowledge of the angiogenesis requirements and the physiology of tissue molding, as well as the development of microsurgical techniques, allowed surgeons to transfer free flaps and prefabricate tissues. A variety of studies have been presented in an attempt to modify or prefabricate the use of our current tissues for discrete defects [8—20]. These methods rely on using donor tissue from patients. The logical progression for plastic surgery is to build upon previous knowledge, combined with tissue engineering, to develop patient-specific, three-dimensional tissue constructs for patient restoration. This article focuses on the applications of tissue engineering to tissue reconstruction following tumor resection. Due to the breadth of the subject matter and, understandably, page limitations, this article is illustrative rather than comprehensive. First, however, the motivation and necessity of tissue engineering in cancer care is presented.

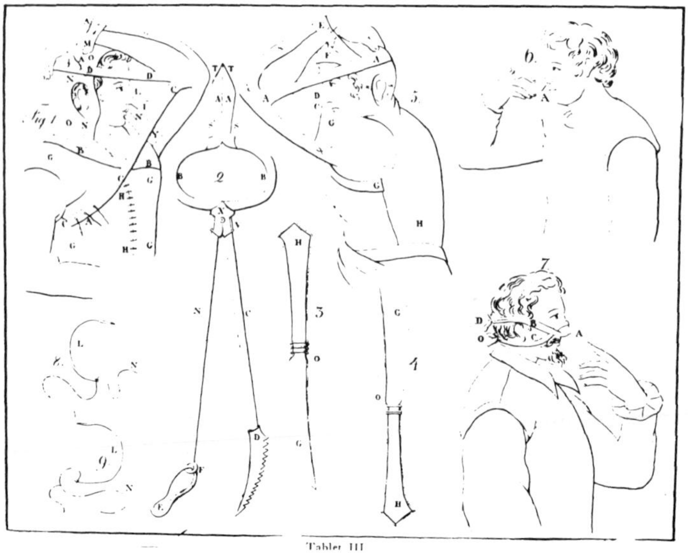

Fig. 1. Tagliacozzi's forearm flap for nasal reconstruction (ca. 1597).

Clinical need for tissue engineering

To fully appreciate the need for tissue engineering in plastic surgery, one must first grasp the goal of tissue reconstruction. Reconstructive surgery in cancer rehabilitation focuses on the repair or restoration of tissues removed following tumor resection so that patient quality of life is maintained or enhanced as measured by resultant tissue function primarily and cosmesis secondarily. Reconstructive techniques must be reliable, expedient, and promote good functional results and cosmesis. The functional goals for cancer reconstruction are to maintain swallowing, prevent aspiration, preserve speech, protect vital structures, allow ambulation, and achieve primary wound healing. Tissue engineered strategies must meet or exceed these goals to be of clinical utility. Whether restoration of contour, i.e., aesthetic cosmesis, is possible currently depends on the defect size, location, and the structures involved.

It is estimated that more than 1 million reconstructive procedures are performed by plastic and reconstructive surgeons every year in the USA to repair tissue defects [21]. Despite our advances, we remain somewhat primitive in our approach. The conventional clinical solution to restoring structure and function of tissue defects relies on "robbing Peter to pay Paul". That is, tissue defects resulting from cancer resection, congenital abnormalities, and trauma are repaired by transferring healthy donor tissue as an autograft, isograft, allograft, or xenograft to a tissue defect. The necessity of donor tissue limits reconstruction and results in certain associated morbidity, e.g., scar formation, potential injury and impairment to the donor site. For instance, mandibular reconstruction may require bone and soft tissue from the scapula or fibula transferred as a free flap. Breast reconstruction robs the abdominal wall of its soft tissue and muscle. Nerve reconstruction may create anesthetic areas in the foot or arm where donor nerves are harvested. Large soft tissue defects may require the use of large back muscles to provide stable coverage. For autografts, the volume of donor tissue that can be harvested is dependent on blood supply and the need to avoid visceral injuries and contour deformities. Allografts and xenografts possess the risks of immune rejection and infectious disease transmission. Although current surgical solutions have resulted in tremendous strides in increasing patient survival and quality of life, the inherent limitations provide an impetus to develop innovative reconstruction strategies.

In addition to the noted shortcomings with conventional reconstructive methods, there is also an economic impetus for developing tissue engineered constructs. Each year in the United States physicians perform an estimated 8 million surgical procedures to treat tissue loss and organ failure, which require more than 40 million days of hospitalization and cost an estimated US$400 billion [6]. Reconstruction following tumor ablation can be quite expensive with prices averaging between US$5,000 and US$17,000. Reconstruction costs can be reduced by decreasing the time required in the operating room (currently charges are approximately US$40/min), the number of hospital days required for patient recovery, patient back-to-work time, and follow-up complications that require additional surgery. Moreover, under the current health care system, insurance companies are becoming reluctant to pay for certain reconstructions as well as follow-up procedures. For instance, 84% of plastic surgeons reported having up to 10 patients who were denied insurance coverage for breast reconstruction following mastectomy in 1995 [21]. The donor scarcity, technical difficulties, expense, and complex labor intensive care associated with conventional tissue reconstruction provides a significant impetus for development of tissue engineered therapies. The field of tissue engineering may allow us to modify tissue or create constructs to supplant donor tissue currently used for reconstruction and, at the same time, provide reparative strategies that curtail the current US$1 trillion/year United States health care costs by reducing operating room time and patient hospital days, as well as deleting the continual need for immunosuppressants.

Reconstructive surgery applications

Driven by the urgent need to develop clinical modalities that will alleviate high health care costs, patient morbidity, donor tissue shortage, and secondary complications realized with contemporary tissue reconstruction procedures, we at the M.D. Anderson Cancer Center and others are focusing on methods to develop tissue engineered adipose, bone, and peripheral nerve, as well as to promote neovascularization to ensure tissue viability. Figure 2 depicts the enormity of a

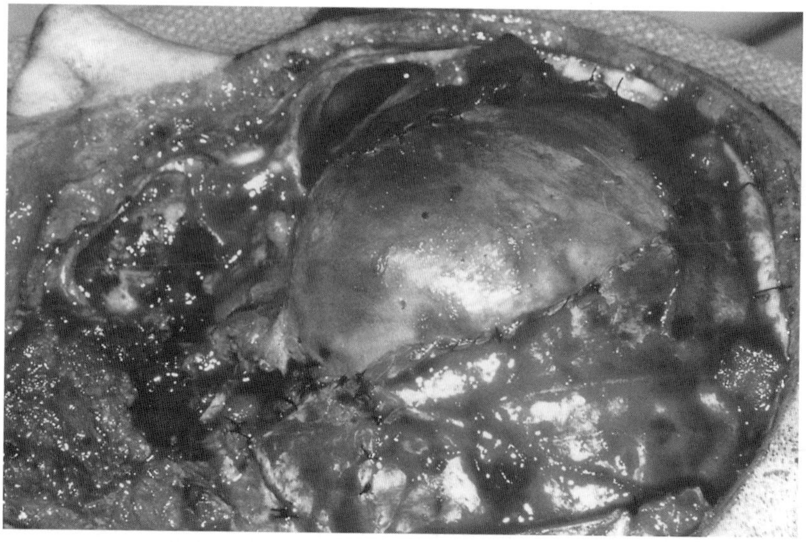

Fig. 2. Patient with a squamous cell carcinoma of the left frontal forehead. Resection proceeded with the resulting defect with exposed dura and the requirements for replacement of bone, nerve, muscle, soft tissue and skin.

tissue defect that follows tumor resection. The patient had a squamous cell carcinoma resected from her left frontal forehead. The defect demonstrates exposed dura with loss of a portion of the cranium, left eye, frontal sinus, external skin, muscle and fat. Reconstructive demands necessitates bone, soft tissue, fat, nerve, and skin. To date, this patient must be reconstructed by "robbing" bone, nerve, cartilage, skin, and adipose from uninjured donor sites, thus creating additional defects within the patient.

Neovascularization

Tissue-engineered epidermal skin and cartilage equivalents have proven successful primarily because these tissues are relatively avascular. That is, the constructs can survive by nutrient diffusion alone. The kinetics and extent of neovascularization determine the fate of survival and subsequent function of more complex tissue-engineered constructs and represents the predominant constraint in current engineering design to date. Tissue engineering modalities have proven successful in various small animal models. However, scale-up of tissue constructs from these animal models to humans is largely hindered by the inability to provide a microvascular network within a construct. This inability impairs mass transport of cellular nutrients and severely limits the clinical utility of constructs designed for human use. Reconstructive surgeons have long realized that the kinetics and extent of the neovascularization process are critical for the survival and subsequent function of free autologous tissue grafts [22]. The same is no less true for tissue-engineered equivalents, which represent the next generation of patient-specific reconstructive tissue grafts.

Tissue growth is naturally a self-regulating system and tissue mass will not increase more than can be adequately supported by diffusion. Volumes of tissue greater than 1—3 μl cannot be implanted into humans because nutrition, gas exchange, and elimination of waste products are limited by diffusion [23]. Folkman demonstrated two decades ago that cells died if they were greater than 150—200 μm from the surface of 1-mm tumor spheroids, resulting in a necrotic core [23]. Others have similarly suggested that the thickness of cell mass must be limited to a few hundred microns to prevent the formation of an anoxic core when designing bioartificial livers [24]. A microvascular network overcomes this problem by matching vascular density with tissue volume.

One method to provide a vascular network is to prevascularize a tissue. Plastic surgeons routinely use this concept to create vascularized flaps. For instance, the greater omentum has been used as a donor tissue in reconstructive surgery for its versatility and reliability and the ability to create vascularized structures after a delay procedure. Erol and Spira demonstrated that omentum tunneled through the abdominal wall and wrapped around cartilage or bone and created a prefabricated flap with a blood supply that could then be transferred to a distant location [17]. Tzuyano has performed similar revascularization studies in rabbits where a severed hind limb was placed in a vascular bundle [13]. The hind limbs within the vascular bundle demonstrated angiogenesis and raised the potential for a possible new method to salvage severed composite tissue segments. Khouri has demonstrated revascularization of rat knee joints which were wrapped with ipsilateral superficial inferior epigastric fascia [8]. Similarly, biodegradable polymer scaffolds can be implanted prior to cell seeding to prevascularize scaffolds. However, this effectively requires two surgical procedures (implanting the scaffold for prevascularization, then removing the scaffold to seed it with cells and place it into its anatomical location). Moreover, fibrovascular tissue often overfills the scaffold, leaving inadequate, void space for cell seeding [25]. Ideally, one would like to incorporate a microvascular network within a tissue engineered construct concomitantly with tissue ingrowth within a scaffold. We are currently defining the neovascularization microenvironment within tissue-engineered constructs by employing novel quantitative methodologies in order to address this goal.

Tissue engineered adipose tissue

In 1995, 182,000 American women were diagnosed with breast cancer and of these, 85,000 patients received a mastectomy as part of their treatment [21]. The goal of breast reconstruction is to restore the breast mound such that it is realistic in feel and appearance, as well as drapes and moves naturally on the chest wall. A total of 25,933 breast reconstruction procedures were reportedly conducted in the United States in 1994 [21]. Reconstruction is conventionally conducted by one of two procedures: tissue expansion followed by breast implant insertion or flap surgery. The former method has been used to such an extent that 1 million American women currently have breast implants [26]. However, the apparent problems noted with silicone gel breast implants led to a US Food and Drug Administration ban on their use in 1992 and the publicity generated has led to a heightened awareness for the critical need for alternative breast reconstruction procedures after mastectomy [26]. Current implant alternatives to silicone gel breast implants are often not satisfactory. Saline-filled implants, which avoid the problems associated with silicone gel, have generated complaints that the implants do not restore similar consistency of the breast. Moreover, the majority of implants have to be removed due to capsular contracture, a process in which fibrous scar tissue forms around the implant as part of the patient's immune response and begins to tighten, causing displacement of the implant, distortion of the breast symmetry, and unrealistic firmness. In 1994, 80% of the 37,853 implants inserted during breast reconstruction were later removed because of capsular contracture [21]. Other realized implant complications include post-operative infection or hematoma.

The current alternative to breast implants is flap surgery. Muscle flap transfers restore similar consistency and permit the transposition of relatively large volumes of skin, underlying fat, and muscle to the breast area. The two most common flaps are the latissimus dorsi musculatocutaneous flap (skin, fat, and muscle from the upper back) and the transverse rectus abdominous musculatocutaneous flap (skin, fat, and muscle from the lower abdominal wall). However, flap procedures are complicated to surgically perform and result in donor site morbidity. Further, they normally require several follow-up procedures.

The emotional distress which accompanies a lump found in the breast is compounded by the apparent problems noted with reconstructive procedures. Without a trusted reconstruction procedure, the specter of a disfiguring operation adds to the fear of the cancer. This fear can lead to patient denial and, inevitably, to avoidance of breast cancer screenings and delayed treatment. Tissue engineered adipose tissue can provide patients with breast tissue comprised of the patient's own cells, negating donor site morbidity and complications associated with flap surgery. New breast reconstruction methods that are reliable, safe, and improve cosmesis can have a positive impact on women, leading to appropriate cancer screening and prompt treatment. Moreover, they can provide a positive psychological benefit for women, giving them a renewed sense of wholeness and a return to a normal state.

Although tissue engineered adipose tissue could be used for soft tissue augmentation, such as in facial augmentation to correct congenital problems, it would have immediate and widespread use in breast reconstruction following tumor resection. The use of autologous tissue source that does not require a significant donor site and avoids the use of a conventional implant would revolutionize reconstructive cancer surgery. For breast reconstruction, we envision the following sequela of events: 1) autologous adipocytes are harvested from a patient via endoscopic liposuction and cultured ex vivo, 2) adipocytes are seeded onto polymer scaffolds of predefined shapes, 3) the cell-seeded scaffolds are implanted onto the patient's pectoralis muscles (two muscles that lie beneath the natural breast tissue), and 4) adipocytes proliferate, the polymer scaffold degrades into cell metabolizable products (glycolic and lactic acid), and the tissue becomes vascularized. As a first step, we are currently culturing adipocytes and seeding them onto polymer scaffolds to assess the cells' lipogenic function and proliferation characteristics. Because adipocytes possess one of the highest metabolic rates in the body, adipose tissue engineering will have to dovetail with the above-mentioned neovascularization methods to be clinically translatable.

Tissue engineered bone

The requirement to repair deficient skeletal locations with bone has been realized for centuries [27,28]. Satisfactory reconstruction of osseous defects requires restoration of stability and length while maintaining mobility. Conventional bone grafting is an option for segmental defects of limited length (<6 cm), but requires stable soft tissue coverage and a well vascularized bed. This method should be avoided in irradiated tissues, which is often the case in cancer patients. Cadaver allografts may be used for large defects that may involve joint surfaces, and heal like conventional bone grafts. Due to their large volume and possible low-grade immune activity, cadaveric allografts may require up to 1 year for healing and are more prone to complications.

Vascularized bone transfer provides a useful alternative to some of these techniques or may be used in conjunction with them to enhance successful bone healing. In this procedure, a segment of bone, e.g., fibula or iliac crest, is harvested with its primary nutrient blood vessels and transferred to a distant site. This technique is limited by donor-site availability, potential added morbidity, and the requirement to shape bone at the time of operation, an often difficult and time consuming task. Tissue engineering vascularized bone into useful shapes and would greatly improve the clinical approach to reconstructive surgery.

Bone healing occurs by a process known as creeping substitution in which a graft is gradually replaced by new bone and blood vessels arising from nearby tissues. Bone grafts promote these events by providing 1) a calcified matrix that serves as a porous osteoconductive scaffold to guide bone growth, 2) a collection of potent cytokine induction factors stored in the noncalcified extracellular matrix, and 3) a small population of osteoblasts and/or progenitor cells that contribute to the bone forming process. This sequela might be accomplished with tissue engineering by using a combination of cell transplantation, cytokine induction factors, and biocompatible polymers. Colleagues have created five-sided tissue molds using poly(methyl methacrylate), filled the molds with morcellized bone graft, and implanted the molds (open side against periosteum) in sheep [29]. Vascularized bone flaps of different shapes could then be transferred to other anatomical locations developed. The new bone persisted for 20 weeks with evidence of maturation. Investigators have seeded polymer scaffolds with osteoblasts in effort to form three-dimensional cell-seeded polymers that can be implanted [30—33]. Osteoblast-seeded polymer templates have been used to reconstruct femoral shafts in rats [34].

Tissue engineered peripheral nerve

The conventional approaches to repairing nerve injuries following tumor resection are either primary nerve repair (suturing together distal and proximal ends of a cut nerve) or autografting (interposing a nerve gap with a donor nerve). In either case, microsurgical techniques are utilized to perform epineural or interfascicular repair. Nerve repair must be tension free. Tension has a direct negative effect on the outcome of nerve regeneration, and if longitudinal gliding is poor, nerve grafting is required. Typical donor nerve sites include cutaneous sensory nerves such as the sural or lateral/medial antebrachial cutaneous nerve. However, this leads to further sensory loss in the donor nerve distribution.

Failure to restore injured nerves can result in the loss of muscle function, impaired sensation and/or painful neuropathies. Each year in the United States, the prolonged recovery from traditionally treated nerve injuries, results in millions of dollars in lost revenue and increased compensation benefits. Our current technology involving nerve repair is limited by the availability of donor tissue and the morbidity related to the sacrifice of donor nerve. The utilization of allografts and immunosuppression as nerve grafts is an option. However, outcomes in nerve reconstruction with the use of allografts have not approached results utilizing autografts. Furthermore, oncologic and traumatic considerations frequently preclude the use of allografts with immunosuppression. A biodegradable bioactive polymer that could serve as a conduit for peripheral nerve regeneration would be superior to what is currently available to us today.

The idea for the development of a nerve conduit is not new. Glück initially utilized decalcified bone as a conduit for nerve regeneration [35]. A variety of other substances have been used including fascia, vein grafts and fallopian tubes. The problem encountered with the use of conduits is their current limitations to small nerve defects. Clinically, nerve defects up to 20 cm may require repair for restoration of function (Fig. 3). From an engineering design standpoint, the ideal conduits should 1) be easily fabricated, 2) be fully biodegradable and replaced by myelinated axons over a several month period, 3) possess low antigenicity, 4) have sufficient porosity to promote vascularization and O_2 transport, 5) be amenable to fabrication into various luminal diameters and lengths, and 6) provide structural integrity and flexibility to resist fracturing upon nerve movement.

Studies to date have concentrated on optimizing conduit polymer chemistry and geometry such that conduits are cyto- and biocompatible, biodegradable, and permissive for nerve regeneration. Table 1 illustrates the wide array of synthetic biodegradable polymers used to fabricate conduits. Previous nerve guides prepared from polyesters, specifically poly (DL-lactic acid), have demonstrated

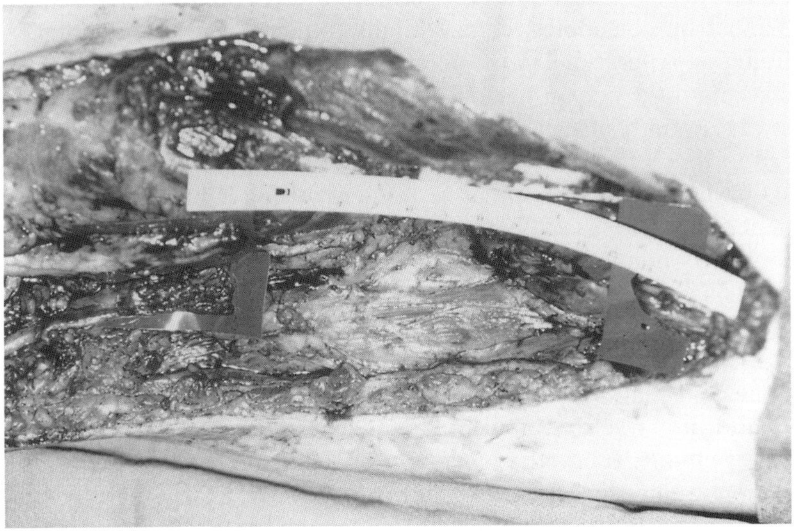

Fig. 3. A 33-year-old male with a 12-cm nerve defect from the sciatic nerve resected for recurrent malignant fibrohistiocytoma. Reconstruction proceeded with use of the sural nerve for autogenous donor nerve grafts.

Table 1. Synthetic, biodegradable nerve conduits.

Polymer configuration	Reference
PGA	[36–39]
GTMC	[40,41]
PLLA	[37,42,43]
PLGA	[44]
PEIP	[45]
PLLA/PCL	[46,47]
PLLA/PCL + PLLA/estane	[48,49]
ANV	[46]
EVA	[50,51]

Abbreviations: PGA = polyglycolic acid; GTMC = glycolide trimethylene carbonate; PLLA = poly-L-lactic acid; PCL = poly-ε-caprolactone; PLGA = poly(lactic-glycolyic) acid; PEIP = poly ((ethylalanato) (midazolyl) posphazene); ANV = acrylonitrile vinylchloride; EVA = ethylene-vinyl acetate.

regeneration. Innervation of the distal nerve stump occurred in conduits, provided the nerve gap was $\leqslant 10$ mm [43]. In another study, bioabsorable polyglycolic acid nerve conduits were compared with the classical sural nerve graft. The bioabsorbable nerve conduit and the sural nerve graft groups had mean nerve fiber diameters, as well as amplitude and conduction velocities significantly less than those of normal control ulnar nerves. Despite the lack of functionality, the nerve conduits guided nerve regeneration [52]. Further studies utilizing poly-L-lactide/poly-caprolactone copolymeric nerve guidance channels and PGA have demonstrated inferior results compared to nerve autografts [39]. These studies emphasize that nerve regeneration and function are distinct but related mechanisms, and both must be considered in developing clinically useful tissue-engineered nerve strategies. Alterations in the geometric configuration of the conduit also appear to have an influence on peripheral nerve regeneration. Synthetic guidance channels with smooth surfaces appear to promote peripheral nerve regeneration compared to rough textures [46,50,51].

Several studies have attempted to enhance this growth by filling the conduits with various materials such as collagen, laminin and fibronectin. Although improved results were obtained over small defects, results still fall short compared to autografts [53,54]. Contrary reports also exist to the efficacy of a matrix-filled conduit. Acrylic copolymer tubes filled with collagen gel or laminin appeared to impede nerve regeneration [46,50,55].

It is becoming evident that Schwann cells play a unique and dynamic role during nerve repair and are largely responsible for the peripheral nervous system's unique ability to regenerate. In axonal regeneration following a peripheral nerve injury or transection, Schwann cells proliferate and migrate from proximal and distal nerve ends and form a continuous tissue bridge across the nerve gap. Schwann cells have been reported to continuously span 10-mm distances in the absence of nerve [55]. Although the exact underlying mechanisms regulating dynamic axon/Schwann cell apposition are unknown, experiments support the concept that Schwann cells offer a highly preferred substrate for axon migration and release bioactive factors which further enhance nerve migration [56,57]. The role of Schwann cells in nerve regeneration can be best illustrated by abrogating Schwann cell function and subsequently accessing nerve regeneration. Rath and colleagues performed an elegant study by constructing a transgene-induced insertional mutation in mice [58]. The mutation, referred to as enervated or Enr, disrupts Schwann cell/axon interactions by impairing the ability of the Schwann cells to differentiate into the correct phenotype. Enr mice demonstrate markedly decreased peripheral nerve regeneration. Hence, Schwann cells play an obligatory role in peripheral nerve regeneration by providing a bioactive substrate on which axons migrate and release bioactive molecules, which regulate axon migration.

In an effort to make conduits bioactive and increase functional outcome of peripheral nerve regeneration, we are presently incorporating Schwann cells and nerve growth factors with variable release kinetics within extruded PLGA conduits. Extruded conduits provide stability and flexibility required for nerve interposition [59]. Briefly, 75:25 PLGA was dissolved in methylene chloride and 150–300 μm salt crystals were added to the polymer solution. The formed suspension was cast in a petri dish, allowed to evaporate, and the resulting PLGA/salt composite disks were cut and placed into a piston extrusion tool that is heated to 250 °C. After heating, the PLGA/salt composite was extruded to form a tube with an inner diameter of 1.6 mm and an outer diameter of 3.2 mm. The tubes were cut to 12 mm and underwent a salt-leaching and vacuum-drying process. These tubular conduits have been interposed into the right sciatic nerve of rats for up to a period of 3 months. All conduits remained flexible allowing mobility of the rat extremity without breakage. Histomorphology demonstrated axonal migration and nerve tissue advancement through the entire conduit and into the distal nerve stump at 3 months. Several axons demonstrated new myelin formation. Walking track analysis revealed restoration of muscle innervation, but no conduction velocity or amplitude was obtained. Elongation and partial collapse of the conduit was noted. This was the first evidence of distal nerve innervation with the use of our fabricated scaffold conduit. We are currently in the process of incorporating Schwann cells and nerve growth factors to increase the rate of nerve regeneration and promote nerve function.

Conclusion

The field of tissue engineering is rapidly expanding. Synergy between bioengineering, life sciences and the clinical science of plastic surgery has preset an interaction which will continue to flourish. Increasing knowledge of cellular and molecular mechanisms and development of sophisticated quantitative tools will culminate in the translation of tissue engineered constructs from the laboratory to the clinic. Tissue engineers Vacanti and Mikos have stated [60],

"The age of "Tissue Engineering" is upon us. As we thrust toward a new millennium, so too are our beliefs that the amalgamation of engineering and medicine will result in an explosion of knowledge that will enhance our understanding of developmental biology and culminate in a new era in medicine enabling us to restore lost tissue function."

It is perhaps premature to predict the beneficial outcome of tissue engineering, but if the innovative ideas being explored and ebullience of collaborating bioengineering, life science, and clinical investigators are any indication, tissue engineering is sure to develop constructs that can be used in reconstruction for patients with tissue defects. Ahead lies a bright future for tissue engineering, and if hurdles do appear, we have only to look to our patients for the motivation to drive forward.

References

1. Habal MB. Perspectives on what's new in plastic surgery: the new millennium. Clin Plastic Surg 1996;23:1–2.
2. Kiki H. Directions in Tissue Engineering: Technologies, Applications, and Regulatory Issues. Inaugural Meeting of the Tissue Engineering Society. Orlando, Florida, 13–15 December, 1996.
3. Tagliacozzi G. De Curtorum Chirurgia per Insitionem. Venice: Gaspare Bindoni, 1597.
4. Desault PJ. Oeuvres Chirurgicales ou Exposé do la Doctrine et de la Plastique. Paris: Megegnon, 1798.
5. Zeis E. Handbuch der plastischen Chirurgie (nebsteiner Vorrede von J.F. Dieffenbach). Berlin: Reimer, 1838.
6. Langer R, Vancanti CA. Tissue engineering. Science 1993;260:920–926.
7. Bhishagratna KK. An English translation of the Sushruta Samhita based on original Sanskrit text. Calcutta: Bose, 1916.
8. Khouri RK, Upton J, Shaw WW. Prefabrication of composite free flaps through staged microvascular transfer: an experimental and clinical study. Plast Recon Surg 1991;87:108–115.
9. Upton J, Ferraro N, Healy G, Khouri R, Merrell C. The use of prefabricated fascial flaps for lining of the oral and nasal cavities. Plast Recon Surg 1994;94:573–579.
10. Özgenta HE, Shenaq S, Spira M. Prefabrication of a secondary TRAM flap. Plast Recon Surg 1995;95:441–449.
11. Itoh Y. An experimental study of prefabricated flaps using silicone sheets, with reference to the vascular patternization process. Ann Plast Surg 1992;28:140–146.
12. Mulliken JB, Glowacki J. Induced osteogenesis for repair and construction in the craniofacial region. Plast Recon Surg 1980;65:553–560.
13. Tzuyao S. Experimental study of tissue graft vascularization by means of vascular implantation and subcutaneous burying. Plast Recon Surg 1984;73:403–410.
14. Erol Ö, Spira M. Utilization of a composite island flap employing omentum in organ reconstruction: an experimental investigation. Plast Recon Surg 1981;68:561–570.
15. Hirase Y, Valauri FA, Buncke HJ. Prefabricated sensate myocutaneous and osteomyocutaneous free flaps: an experimental model. Preliminary report. Plast Recon Surg 1988;82:440–445.
16. Stark GB, Hong C, Futrell JW. Enhanced neovascularization of rat tubed pedicle flaps with low perfusion of the wound margin. Plast Recon Surg 1987;80:814–824.
17. Khouri RK, Koudsi B, Reddi H. Tissue transformation into bone in vivo. JAMA 1991;266: 1953–1955.
18. Khouri RK, Tark KC, Shaw WW. Prefabrication of flaps using an arteriovenous bundle and

angiogenesis factors. Surg Forum 1992:597–599.
19. Hussl H, Russell RC, Zook EG, Eriksson E. Experimental evaluation of tissue revascularization using a transferred muscular-vascular pedicle. Ann Plast Surg 1986;17:299–305.
20. Hyakusoku H, Okubo M, Umeda Ta, Fumiiri M. A prefabricated hair-bearing island flap for lip reconstruction. Br J Plast Surg 1987;40:37–39.
21. American Society of Plastic and Reconstructive Surgeons (ASPRS) and Plastic Surgery Educational Foundation (PSEF). Data collection by these two societies. Database Survey, 1995.
22. Höckel M, Schlenger K, Frischmann-Berger R, Berger S, Vaupel P. Therapeutic angiogenesis in surgery and oncology. In: Maragoudakis ME (ed) Angiogenesis: Molecular Biology, Clinical Aspects. New York: Plenum Press, 1994;193–207.
23. Folkman J, Hochberg M. Self-regulation of growth in three dimensions. J Exp Med 1973;138: 745–753.
24. Berthiaume F, Yarmush ML. Tisue engineering. In: Bronzino JD (ed) The Biomedical Engineering Handbook. Boca Raton: CRC Press, 1995;1556–1577.
25. Wake MC, Patrick Jr. CW, Mikos AG. Pore morphology effects on the fibrovascular tissue growth in porous polymer substrates. Cell Transpl 1994;3:339–343.
26. Angell M. Evaluating the health risks of breast implants: the interplay of medical science, law, and public opinion. N Engl J Med 1996;334:1513–1518.
27. Senn N. On the healing of aseptic cavities by implantation of antiseptic decalcified bone. Am J Med Sci 1889;98:219–243.
28. Meek'ren JJV. Observations medico-chirurgicae. Amsterdam: H and T Boon, 1632.
29. Miller MJ, Goldberg DP, Yasko AW, Lemon JC, Satterfield WC, Wake MC, Mikos AG. Guided bone growth in sheep: a model for tissue-engineered bone flaps. Tis Eng 1996;2:51–59.
30. Ishaug-Riley SL, Crane GM, Gurlek A, Miller MJ, Yaszemski MJ, Yasko AW, Mikos AG. Ectopic bone formation by marrow stromal osteoblast transplantation using poly(DL-lactic-co-glycolic acid) foams implanted into the rat mesentery. J Biomed Mater Res (In press).
31. Ishaug SL, Crane GM, Miller MJ, Yasko AW, Yaszemski MJ, Mikos AG. Bone formation by three-dimensional stromal osteoblast culture in biodegradable polymer scaffolds. J Biomed Mater Res (In press).
32. Laurencin CT, Attawia MA, Elgendy HE, Herbery KM. Tissue engineered bone-regeneration using degradable polymers: the formation of mineralized matrices. Bone 1996;19(Suppl 1): S93–S99.
33. Rivard CH, Chaput C, Selmani A. Bioabsorbable synthetic polyesters and tissue regeneration: a study on the three-dimensional proliferation of ovine chondrocytes and osteoblasts. Ann Chir 1996;50:651–658.
34. Puelacher WC, Vacanti JP, Ferraro NF, Schloo B, Vacanti CA. Femoral shaft reconstruction using tissue engineered growth of bone. Int J Oral Maxillofac Surg 1996;25:223–238.
35. Glück T. Ueber neuroplastik aut dem Wege der Transplantation. Arch Klin Chir 1880;25: 606–616.
36. Henry E, Chiu TH, Nyilas E, Brushart TM, Dikkes P, Sidman RL. Nerve regeneration through biodegradable polyester tubes. Exp Neurol 1985;90:652–676.
37. Kiyotani T, Nakamura T, Shimizu Y, Endo K. Experimental study of nerve regeneration in a biodegradable tube made from collagen and polyglycolic acid. ASAIO J 1995;41:M657–M661.
38. Mackinnon SE, Dellon AL, O'Brien JP. Changes in nerve fiber numbers distal to a nerve repair in the rat sciatic nerve model. Muscle Nerve 1991;14:1116.
39. Rosen JM, Pham HN, Abraham G, Harold L, Hentz VR. Artificial nerve graft compared to autograft in a rat model. J Rehab Res 1989;26:1–360.
40. Keeley R, Atagi T, Sabelman E, Padilla J, Kadlcik S, Keeley A, Nguyen K, Rosen J. Peripheral nerve regeneration across 14-mm gaps: a comparison of autograft and entubulation repair methods in the rat. J Recon Microsurg 1993;9:349–358.
41. Rosen JM, Padilla JA, Nguyen KD, Siedman J, Pham HN. Artificial nerve graft using glycolide trimethylene carbonate as a nerve conduit filled with collagen compared to sutured autograft

in a rat model. J Rehab Res 1992;29:1–12.
42. Madison RD, Silva C, Dikkes P, Sidman RL, Chiu TH. Peripheral nerve regeneration with entubulation repair: comparison of biodegradable nerve guides versus polyethylene tubes and the effects of a laminin-containing gel. Exp Neurol 1987;95:378–390.
43. Seckel BR, Chiu TH, Nyilas E, Sidman RL. Nerve regeneration through synthetic biodegradable nerve guides: regulation by the target organ. Plast Recon Surg 1984;14:173–181.
44. Reid RL, Cutright DE, Garrison JS. Biodegradable cuff an adjunct to peripheral nerve repair: a study in dogs. Hand 1978;10:259–266.
45. Langone F, Lora S, Veronese FM, Caliceti P, Parnigotto PP, Valenti F, Palma G. Peripheral nerve repair using a poly(organo)phosphazene tubular prosthesis. Biomaterials 1995;16:347–353.
46. Aebischer P, Guénard V, Valentini RF. The morphology of regenerating peripheral nerves is modulated by the surface microgeometry of polymeric guidance channels. Brain Res 1990;531: 211–218.
47. Dunnen WFA, Lei B, Schakenraad JM, Blaauw EH, Stokroos I, Pennings AJ, Robinson PH. Long-term evaluation of nerve regeneration in a biodegradable nerve guide. Microsurgery 1993;14:508–515.
48. Dunnen WFA, Lei B, Robinson PH, Holwerda A, Pennings AJ, Schakenraad JM. Biological performance of a degradable poly(lactic acid-e-caprolactone) nerve guide: influence of tube dimensions. J Biomed Mater Res 1995;29:757–766.
49. Hoppen HJ, Leenslag JW, Pennings AJ, Lei B, Robinson PH. Two-ply biodegradable nerve guide: basic aspects of design, construction, and biological performance. Biomaterials 1990; 11:286–290.
50. Aebischer P, Salessiotis AN, Winn SR. Basic fibroblast growth factor released from synthetic guidance channels facilitates peripheral nerve regeneration across long nerve gaps. J Neurosci Res 1989;23:282–289.
51. Guénard V, Dinarello CA, Weston PJ, Aebischer P. Peripheral nerve regeneration is impeded by interleukin-1 receptor antagonist released from a polymeric guidance channel. J Neurosci Res 1991;29:396–400.
52. Dellon AL, Mackinnon SE. An alternative to the classical nerve graft for the management of the short nerve gap. Plast Recon Surg 1988;82:849–856.
53. Keeley R, Atagi T, Sabelman E, Padilla J, Kadlcik S, Keeley A, Nguyen K, Rosen J. Peripheral nerve regeneration across 14 mm gaps: a comparison of autograft and entubulation repair methods in the rat. J Recon Microsurg 1993;9:349–360.
54. Tong XJ, Hirai KI, Shimada H, Mitzutani Y, Izumi T, Toda N, P. Yu P. Sciatic nerve regeneration through synthetic biodegradable nerve guides: regulation by the target organ. Brain Res 1994; 663:155–162.
55. Madison RD. Point sources of Schwann cells result in growth into nerve entubulation repair site in the absence of axons: effects of freeze-thawing. Exp Neurol 1994;128:266–275.
56. Tonge DA, Golding JP. Regeneration and repair of the peripheral nervous system. Sem Neurosci 1993;5:385–390.
57. Martini RJ. Expression and functional roles of neural cell surface molecules and extracellular matrix components during development and regeneration of peripheral nerve. Neurocytol 1994;3:1–28.
58. Rath EM, Kelly D, Bouldin TW, Popko ZB. Impaired peripheral nerve regeneration in a mutant strain of mice (Enr) with a Schwann cell defect. J Neurosci 1995;15:7228–7237.
59. Widmer MS, Evans GRD, Brandt K, Savel T, Patrick CW, Mikos AG. Manufacture of Porous Biodegradable Polymer Scaffolds for Nerve Regeneration. 15th Annual Houston Conference on Biomedical Engineering Research. Houston, Texas, 13–14 February 1997.
60. Vacanti CA, Mikos AG. Letter from the editors. Tis Eng 1995;1:1–2.

Tissue Engineering for Therapeutic Use 1.
Y. Ikada and Y. Yamaoka, editors.

Bone regeneration by differentiation factors (BMPs)

Shinichiro Oida
Department of Biochemistry, Faculty of Dentistry, Tokyo Medical and Dental University, Yushima, Tokyo, Japan

Abstract. Initiation of bone formation from undifferentiated mesenchymal cells is an important problem for postnatal bone regeneration, as well as embryonic bone formation. Bone morphogenetic proteins (BMPs) are significant initiators of chondrogenesis and osteogenesis. BMPs, originally identified in the extract from demineralized bone matrix, induce cartilage and bone formation by their intramuscular or subcutaneous implantation. Several BMP genes, members of the TGF-β superfamily, were isolated, and recombinant human BMPs were prepared using gene expression systems. Biological activity of recombinant BMPs has been investigated in vitro and in vivo assay systems, and the application of recombinant human BMPs to therapeutic use has been considered.

Introduction

The discovery of bone morphogenetic proteins (BMPs) came from a study of ectopic cartilage and bone formation by demineralized bone implantation into nonosteogenic sites. BMPs are accumulated in the matrix of bone and tooth dentin and are thought to initiate the regeneration of these hard tissues. Attempts were made to purify BMPs from bovine bone using ectopic bone formation as the assay system. Several bone inductive proteins were finally identified by cDNA cloning, and the activity was determined using recombinant proteins [1].

BMPs are all secreted proteins consisting of signal sequences, followed by propeptides and mature regions (Fig. 1A). Biologically active BMPs are a dimeric protein of mature regions (Fig. 1A) and belong to the TGF-β superfamily based on the primary amino acid sequence of the molecule. Figure 1B shows a TGF-β superfamily member detected in bone tissue. GDF is another TGF-β subgroup closely related to the BMP family, and GDF-5 and GDF-6 also exhibit osteoinductive activity (Fig. 1B).

Purification from bovine bone matrix

There are two methods for preparing osteoinductive BMP. One method is to purify BMP from bone extracts, and the other is to produce the recombinant protein using the BMP gene expression system. In the purification method from the bone matrix, bovine bone powder was decalcified with 0.5 M hydrochloric acid

Address for correspondence: Shinichiro Oida, Department of Biochemistry, Faculty of Dentistry, Tokyo Medical and Dental University, 1-5-45 Yushima, Bunkyo-Ku, Tokyo 113-8549, Japan. Tel.: +81-3-5803-5450. Fax: +81-3-5803-0187. E-mail: s.oida.obch@dent.tmd.ac.jp

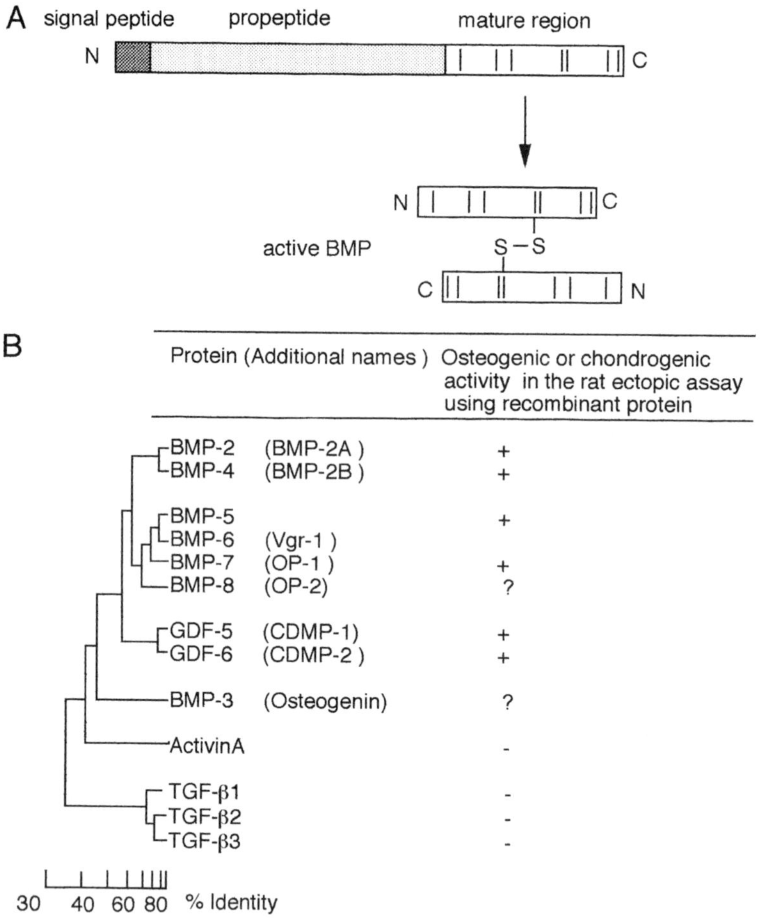

Fig. 1. BMP structure and the TGF-β superfamily identified in bone. **A:** General precursor structure of the TGF-β superfamily proteins containing BMP is shown schematically with the location of the seven conserved cysteine (vertical bars) in the mature region. An active BMP is a disulfide-linked dimeric protein of the mature regions. **B:** The TGF-β superfamily proteins identified in bone and their osteoinductive activity determined by the rat ectopic assay using recombinant proteins.

and then extracted with 4 M guanidine-hydrochloride. The extract was further purified by a series of column chromatographic procedures including gel filtration, ion-exchange, heparin affinity and reversed phase-column chromatography. SDS-polyacrylamide gel analysis of a final product purified from the bone extracts based on the bone inductive activity indicated a broad band of about 35 kDa protein under nonreducing conditions. When the protein was reduced, two components at about 16 and 18 kDa were detected. Amino acid sequence analysis indicated that BMP-3 (osteogenin) was the major BMP component in our final product, but the osteoinductive activity of BMP-3 has not yet been determined by the recombinant protein. The highly purified osteoinductive fraction is an effective material for bone induction, but a mixture of BMPs. It is difficult to isolate a single BMP from the mixture containing several BMP proteins (Fig. 1B).

Production of recombinant BMP

To obtain a large quantity of individual BMP, the recombinant protein expression system is a useful method. Several expression systems using *Escherichia coli*, yeast, plant cells, mammalian cells and baculovirus are available for the production of recombinant protein. It has been reported that TGF-β superfamily proteins are produced by *E. coli*, mammalian cells and baculovirus systems. We have examined these three expression systems. The results indicated that the highest level of active BMPs was expressed in the baculovirus system.

To obtain BMP cDNAs for producing recombinant BMP by the baculovirus system, we prepared mRNA from the cultured human dental pulp cells and constructed a cDNA library. Several BMP genes were cloned from the cDNA library of human dental pulp cells and identified as BMP-2, -4, -6 cDNAs by nucleotide sequence analysis [2]. The clones of BMP-2 and -4 contained the complete coding sequence. Full length BMP-2 and -4 were subcloned into transfer vectors for the baculovirus system [3]. Baculovirus has a strong promoter for the polyhedrin protein, and the wild type virus induces a large amount of polyhedrin protein using this promoter in the infected cells. When polyhedrin and BMP genes are exchanged, the recombinant baculovirus induce a lot of BMP protein using a polyhedrin promoter. Biological activity of recombinant BMP-2 and -4 was determined by in vitro and in vivo assay systems. The recombinant BMPs induced alkaline phosphatase activity in the bone-marrow-derived cell line (ST2) (Fig. 2). Alkaline phosphatase is an osteoblastic maker protein in the early differentiation stage. The recombinant BMPs also initiated endochondral bone formation in the subcutaneous implantation in rats (Fig. 3).

Histology of the cells induced by BMP

BMP is reconstituted with collagen as a carrier matrix. Collagen is a common fibrous protein in the animal body. When the pellets are implanted on the muscle layer under the skin on the back of the rat, BMP-target cells move into the pellet, and differentiate into cartilage, bone and bone marrow cells. We can observe, after 3 weeks, that an originally white and soft pellet has changed into a red and hard pellet. In soft X-ray radiographs, radiopaque material indicating bone formation was observed in the BMP reconstituted pellets.

The process of ectopic bone formation was observed histologically. Undifferentiated mesenchymal cells migrated into the implanted pellet and began to differentiate into chondrocytes. Almost all the cells in the pellet become mature chondrocytes 7 days after implantation, which are easily detected by metachromatic staining with toluidine blue (Fig. 3B). The mature chondrocytes and implanted

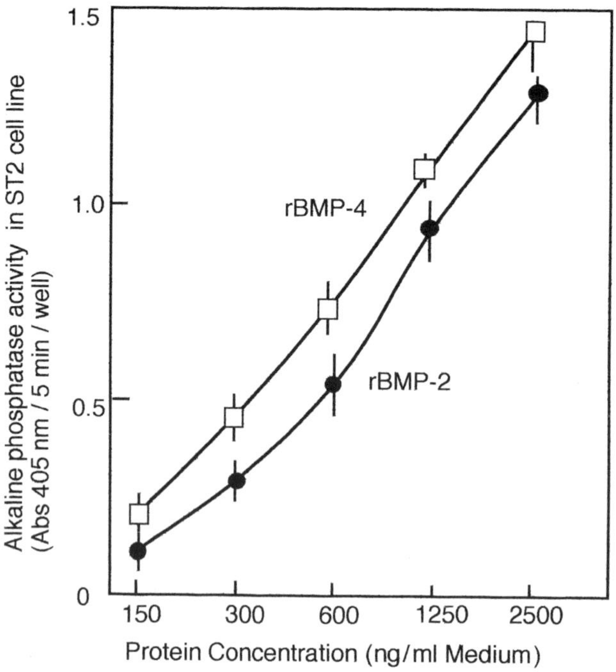

Fig. 2. Biological activity of recombinant BMPs produced by baculovirus system. Recombinant BMPs from insect cells are prepared at the concentration of 1 mg/ml in PBS solution and applied to cultured ST-2 cells.

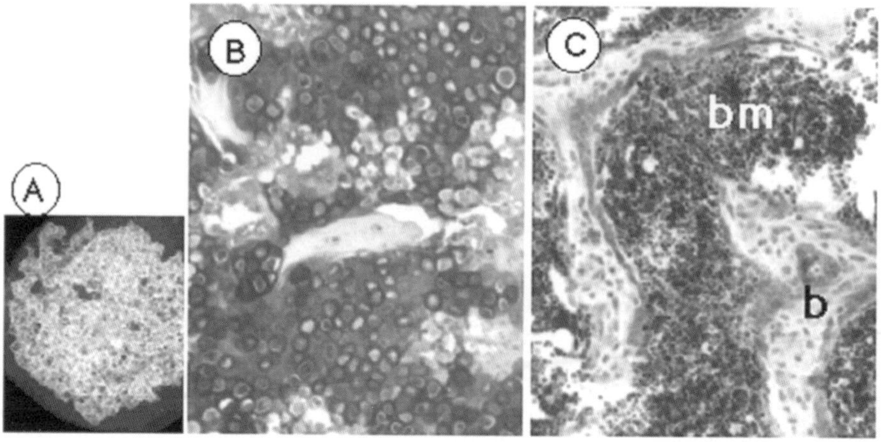

Fig. 3. Soft X-ray radiograph 3 weeks after implantation (**A**) and histology of the cells in the BMP implanted pellet. Cartilage formation is observed 7 days after implantation (**B**) and bone and bone marrow cells 14 days after implantation (**C**). Same differentiation process are observed in bone-derived and recombinant BMP implantation. b = bone tissue, bm = bone marrow cells.

collagen matrix have disappeared 14 days after implantation and were replaced by bone and bone marrow cells (Fig. 3C). By the 3rd week after implantation, mature bone formation was complete. Hard and compact bone was formed in the outer region of the implanted pellet and sponge bone and mature fatty bone marrow tissue was observed inside.

There are two distinct processes of embryonic bone formation. Several bones in a head are generated directly during a process of intramembranous bone formation. However, long bones in a body are generated during a process of endochondral bone formation. In the endochondral bone formation, cartilage is formed first and then replaced by bone and bone marrow cells. The role of BMP in postnatal bone regeneration is that it initiates the differentiation cascade of endochondral bone formation.

BMP receptors

Figure 4A shows the schematic structures of type I and type II receptors for BMPs. A complex of the type I and type II receptors mediates the BMP signal to the intracellular signaling pathway. Both type I and type II receptors have serine/threonine kinase domain inside the cells. Three kinds of type I receptors (BMPR-IA, BMPR-IB and ActR-IA) and one type II receptor for BMPs have been reported for mediating BMP signaling [4] (Fig. 4B). We examined the expression of type I receptors by the RT-PCR method during BMP-induced bone formation. Only BMPR-IB was expressed in the skeletal muscle cells before BMP implantation. The cascade of endochondral bone formation may begin with this type I receptor in skeletal muscle. After BMP implantation, mRNAs of BMPR-IA, BMPR-IB and ActR-IA were expressed at high levels throughout the stages of BMP-induced bone formation. These type I receptors seem to be involved in this BMP-induced bone formation (Table 1).

Homeobox gene expression

Homeobox genes are thought to encode DNA binding transcription factors and play important roles in embryogenesis. The expression of clustertype homeobox gene was detected in skeletal muscle and the cells migrated into the implanted pellet at the early stage by the RT-PCR method (Table 1), but the gene expression was rapidly down regulated and another type of homeobox gene, Msx, was detected 6 and 9 days after implantation. Nucleotide sequence analysis indicated that muscle cells expressed Hox A10, C4 and C9 cluster genes, and invaded cells 3 days after implantation expressed Hox A2, A5, B6, B7, B8, B9, C4, C6 and C9. Msx-2 was expressed in the cells 6 and 9 days after implantation, and Msx-1 was expressed in the cells 9 days after implantation [5]. Similar changes in homeobox genes have been reported in embryonic bone formation. The homeobox gene expression may be an essential molecular event in postnatal bone regeneration as well as embryonic bone formation.

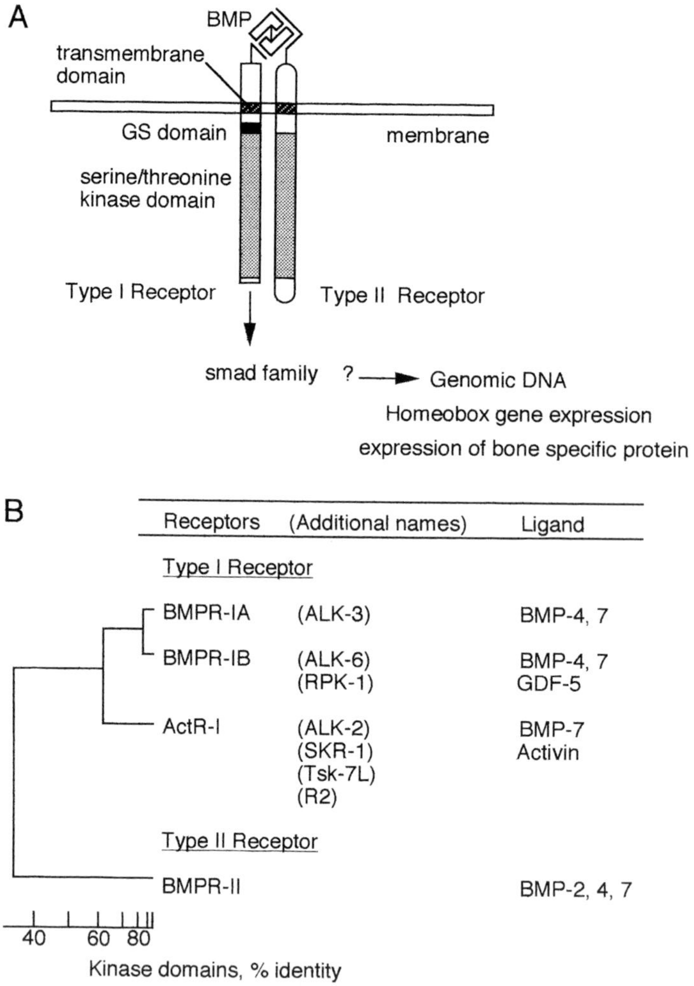

Fig. 4. Structures of BMP receptors and the receptor family. **A:** Schematic structure of type I and type II receptors for BMP. The type I and type II receptor complex mediates the BMP signal to the intracellular signaling pathway. **B:** Serine/threonine kinase receptors mediated BMP signaling, and the known ligand of type I and type II receptor subfamilies is shown.

Target cells

To identify the origin of cells responding to BMP in the early stage, a BMP pellet was implanted with a membrane filter into a subcutaneous region. Cell migration was interrupted by the placement of a membrane filter at the cutaneous or muscle side. When the cell migration was interrupted at the cutaneous side, bone formation was observed in the implanted pellet 3 weeks after implantation as usual. When the cell migration was interrupted at the muscle side, there was no bone

Table 1. Expression of BMP receptors and homeobox genes in ectopic bone formation.

	Skeletal muscle	Cells in the implanted BMP pellet		
		3 days	6 days	9 days
Histology		Fibroblastic mesenchymal cells	Chondrocytes	Bone cells bone marrow cells
Gene expression				
BMP receptor				
BMPR-1A	–	++	++	++
BMPR-1B	++	++	+	+
ActR-1A	–	++	+	+
Homeobox				
HOX (Cluster)	+	+	–	–
Msx-1	–	–	–	+
Msx-2	–	–	+	+

formation in the implanted BMP pellet, however, a bone formation was observed on the filter at the muscle side. The result suggests that the cells from the muscle side initiated the bone formation.

To investigate the reaction of target cells to BMPs, several kinds of cells were enclosed in a plastic ring and membrane filter (diffusion chamber) with the BMP pellet and implanted intraperitoneally. Serum and oxygen pass through the filter, but cells cannot move thorough the filter. In the results of these diffusion chamber experiments, there were no osteogenic cells in the implantation of subcutaneous fibroblastic cells (Table 2). Only cartilage tissue was observed in the skeletal muscle cell 3 weeks after implantation. However, when the cultured skeletal muscle cells were used, neither cartilage nor bone were detected (Table 2). Bone marrow tissue and cultured bone marrow cells differentiated into both cartilage and bone cells [6]. BMPs are thought to induce undifferentiated mesenchymal cells to cartilage and bone and also stimulate osteogenic processes in osteoprogenitor cells. Bone marrow cells consist of undifferentiated mesenchy-

Table 2. Differentiation of mesenchymal cells to BMPs in diffusion chamber assay.

Implanted cells	Induced cells in diffusion chamber
Connective tissue cells	–
Cultured cells	–
Muscle cells	Cartilage
cultured cells	–
Bone marrow cells	Cartilage and bone
Cultured cells	Cartilage and bone

mal cells and differentiated osteoprogenitor cells. These cells in the bone marrow tissue strongly react to BMPs, and bone marrow cells can be effectively used with BMPs for bone regeneration in clinical use.

Studies for the future

Postnatal bone regeneration is induced by BMP in a process such as embryonic endochondral bone formation. In this process, a cellular cascade of programmed cells is initiated by BMPs to osteogenesis. There are three important factors in this initiation. Bone morphogenetic proteins (BMPs) are important as differentiation or initiation factors. Both target cells and carrier matrix are also important for BMP-induced endochondral bone formation. An appropriate matrix is needed for the reaction field of BMP and the target cells. The shape of the matrix decides the shape of the induced bone. Progress in the studies of these three important factors will enable us to realize our dream of regenerating human bone tissue for therapeutic use.

Acknowledgements

Many people at Tokyo Medical and Dental University have contributed to this work. I would like in particular to acknowledge Dr Ashahina and Dr Harada for the histological analysis and animal experiments, and Dr Maruoka, Dr Iimura and Dr Takeda for the study by molecular biological experiments.

References

1. Wozney JM. The bone morphogenetic protein family and osteogenesis. Molec Reprod Devel 1992;32:160–167.
2. Takeda K, Oida S, Goseki M, Iimura T, Maruoka Y, Amagasa T, Sasaki S. Expression of bone morphogenetic protein gene in the human dental pulp cells. Bone 1994;15:467–470.
3. Maruoka Y, Oida S, Iimura T, Takeda K, Asahina I, Enomoto S, Sasaki S. Production of functional human bone morphogenetic protein-2 using a baculovirus/Sf-9 insect cell system. Biochem Mol Biol Int 1995;35:957–963.
4. Massague J, Attisano L, Warana JL. The TGF-β family and its composite receptors. Trends Cell Biol 1994;4:172–178.
5. Iimura T, Oida S, Takeda K, Maruoka Y Sasaki S. Changes in homeobox-containing gene expression during ectopic bone formation induced by bone morphogenetic protein. Biochem Biophys Res Commun 1994;201:980–987.
6. Harada K, Oida S, Sasaki S. Condrogenesis and osteogenesis of bone marrow-derived cells by bone-inductive factors. Bone 1998;9:177–183.

Tissue Engineering for Therapeutic Use 1.
Y. Ikada and Y. Yamaoka, editors.

A gel-entrapment hollow fiber bioartificial liver employing xenogeneic hepatocytes

Wei-Shou Hu[1], Julie R. Friend[1], Florence J. Wu[1], Rory P. Remmel[2] and Frank B. Cerra[3]
Departments of [1]Chemical Engineering and Materials Science, [2]Medicinal Chemistry, and [3]Surgery, University of Minnesota, Minneapolis, USA

Abstract. A bioartificial liver (BAL) employing xenogeneic hepatocytes in a hollow fiber bioreactor was developed by a research team at the University of Minnesota. Hepatocytes entrapped in collagen were loaded in the luminal side of ultrafiltration hollow fibers. Within this entrapment system, hepatocytes were shown to sustain liver-specific metabolic activities of biosynthesis, biotransformation and conjugation. Using an anhepatic rabbit model, rat hepatocytes were shown to continue to function after the BAL was linked to the test animal. Subsequently, a D-galactosamine overdose canine liver failure model was used for scale-up studies. In order to procure a sufficient number of hepatocytes for large animal treatment, a collagenase perfusion protocol was established for harvesting porcine hepatocytes at high yield and viability. An improved survival of dogs treated with the BAL was shown over the controls. For anticipated clinical applications, we explored the possibility of employing hepatocyte spheroids to increase the liver-specific activities in the BAL rather than further increasing the reactor volume. Spheroids are self-assembled tissue-like hepatocyte aggregates formed from monolayer cultures. They exhibit higher liver-specific functions and remain viable longer than hepatocytes in a monolayer. In order to form spheroids more easily on a larger scale, we developed a method of forming spheroids in stirred tank bioreactors for both rat and porcine hepatocytes. The spheroids formed in stirred tanks were morphologically and functionally indistinguishable from those formed from a monolayer. Liver-specific functions of collagen-entrapped spheroids were maintained at higher levels even longer than those of spheroids in suspension. For use in the BAL, a mixture of spheroids and dispersed hepatocytes was used to ensure a proper degree of collagen contraction. A BAL employing a mixture of spheroids and dispersed cells was shown to sustain higher levels of liver-specific functions. The feasibility of using a BAL entrapped with hepatocyte spheroids as a temporary support of liver failure patients warrants further studies.

Keywords: bioartificial liver, hepatocyte spheroids, liver metabolism, self-assembly.

Introduction

Over 30,000 people die of liver failure every year in the USA alone [1]. Mortality rates are particularly high for those diagnosed with acute fulminant hepatic failure (FHF) because hepatic regeneration is neither rapid nor sufficient enough to sustain the patient [2]. Many patients in hepatic failure do not qualify for transplantation because of concurrent medical problems, such as metastatic cancer,

Address for correspondence: Wei-Shou Hu, Department of Chemical Engineering and Materials Science, University of Minnesota, 421 Washington Ave. SE, MN 55455-0132, USA. Tel.: +1-612-625-0546. Fax: +1-612-626-7246. E-mail: wshu@cems.umn.edu active alcoholism, or concomitant infection. Those who do qualify for a liver transplant often die while awaiting an allograft because of the scarcity of donor organs [3]. Thus there is a critical need for improved temporary liver support for potential transplant recipients, as well as for patients with reversible, acute hepatitis who do not qualify for liver transplantation [4].

During the past 40 years, several mechanical techniques for artificial liver support have been attempted for treatment of hepatic failure. These include hemodialysis for the removal of ammonia [5], as well as hemoperfusion using a direct cation exchanger (Dowex 50-X8) [6], hemoperfusion through albumin coated Amberlite XAD-7 resin [7], and hemodialysis across polyacrylonitrile (PAN) membranes [8] for the removal of toxic metabolites which build up in the plasma during liver failure. None of these methods proved to be sufficient for treating liver failure.

For this reason, addition of biological components to the treatment was considered necessary by most investigators, thus leading to the emergence of hybrid "bioartificial" liver systems. Development of these systems, containing biological components within a synthetic framework, was propelled by the advent of extracorporeal perfusion against whole livers [9,10] and liver tissue [11], and the use of isolated hepatocytes [12,13]. These bioartificial liver systems provide greater numbers of liver functions than cellular components or isolated enzyme systems since they utilize intact, metabolically active liver cells [14].

Hepatocyte entrapment bioartifical liver (BAL)

The University of Minnesota BAL device is unique in that cells are entrapped within the lumen of the hollow fibers rather than in the extracapillary space [15,16] (Fig. 1). Hepatocytes harvested by a collagenase perfusion method are suspended in type I collagen solution at 4 °C, and this hepatocyte-collagen suspension is injected into the hollow fiber lumen. Cell culture medium at 37 °C is then perfused in the extracapillary space to raise the temperature of the hepatocyte-collagen suspension. The collagen gels in approximately 15 min, entrapping the hepatocytes within the lumen space. For the following 24 h, culture medium is continuously perfused through the extracapillary space to supply nutrients and oxygen to the hepatocytes. The collagen gel contracts to 60% of its original diameter during this time. Following this initial period, intraluminal perfusion is begun in order to supply hepatocytes with the nutrients and growth factors required to sustain their function, and the bioartificial liver is connected to the test animal in liver failure. The patient's blood is continuously circulated through the extracapillary space. Toxic components in the patient's blood can diffuse through the hollow fiber membrane into the luminal space where they can be metabolized by the entrapped hepatocytes. The detoxified metabolites can then either diffuse back into the patient's blood stream or exit with the intraluminal stream. To separate the intraluminal from the extracapillary space, a semipermeable membrane with a molecular weight cut off (MWCO) of 100,000 Da is used

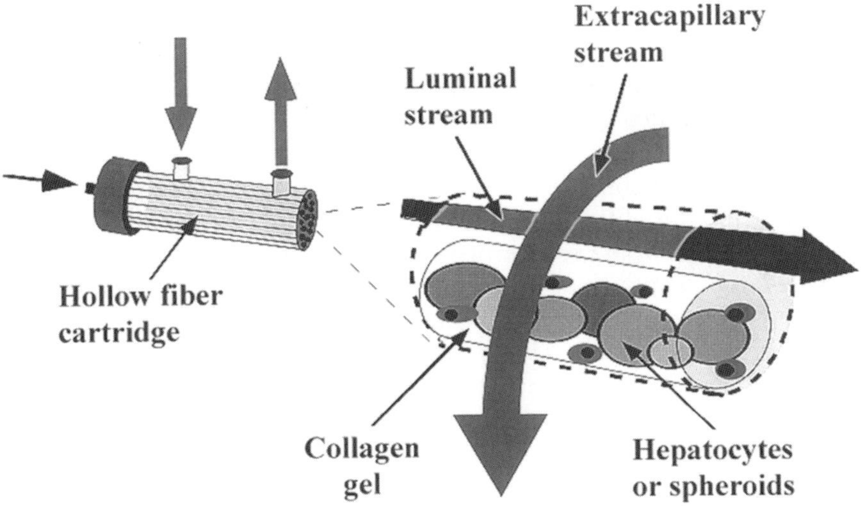

Fig. 1. Schematic diagram of a gel-entrapment hollow fiber bioreactor as a bioartificial liver. For animal trials, unaggregated hepatocytes were entrapped within the gel. To improve device performance, a mixture of spheroids and dispersed hepatocytes may be entrapped in the hollow fiber bioreactor.

in the BAL. The membrane excludes the patient's immunoglobulin and complement proteins from the intraluminal space, while allowing free diffusion of serum albumin and small molecular weight compounds.

Application of the BAL in animal models

In initial in vitro studies, oxygen consumption, albumin biosynthesis, lidocaine clearance and bilirubin conjugation by rat hepatocytes entrapped in a BAL were evaluated for over 7 days of culture [17,18]. Upon demonstration of these liver-specific functions by the entrapped hepatocytes, an anhepatic rabbit model was developed to evaluate the in vivo function of the BAL [19,20]. Rat hepatocytes were inoculated into the bioreactor, and a medium stream was recirculated in the extracapillary space. Over the next 22 h, hepatocyte viability was monitored by measuring oxygen consumption, while the collagen gel contracted to create space for an intraluminal flow stream. After this time, a total hepatectomy was performed on the test animal. Medium was drained from the extracapillary circuit of the BAL, which was then connected to the animal by using an artereovenous shunt. The animal's blood was perfused through the extracapillary space of the BAL at 35 ml/min, and hepatocyte-supporting medium was perfused intraluminally at 9 ml/h. A total of 18 rabbits were divided into three treatment groups: 1) no BAL control, 2) BAL without cells, and 3) BAL with entrapped hepatocytes. Significant reversal of liver failure was observed in the group of animals which received treatment with the hepatocyte-BAL. This improvement was exhibited by a measured decrease in the levels of toxic metabolites in the jaundiced animal plasma.

As a step toward scale-up for human trials, a large animal liver failure model was developed. Liver failure was induced in canines by administering D-galactosamine at a dosage which caused fatality within 48 h [21]. Relevant clinical parameters, including intracerebral pressure (ICP) and assessment of comotosis, were established. Because of the large number of cells needed for treatment, rats were no longer an appropriate source of hepatocytes. Instead, 10-kg pigs were used to provide large quantities of hepatocytes from one source. A sequential in situ portal vein perfusion method using ethylene glycinetetraacetic acid and 0.1% collagenase was developed [22]. In a typical pig liver perfusion procedure, 19.5 billion cells with 94% viability were routinely obtained.

Because of the large number of cells needed for canine trials, the hollow fiber bioreactor used previously in rabbit trials was no longer adequate. A number of hollow fiber bioreactors with a designated molecular weight cutoff of 100,000 Da were evaluated for permeability to serum albumin, the leakage of collagen prior to its gelation, and the passage of immunoglobulin to the luminal side from the shell side. A cellulose diacetate-based hollow fiber bioreactor was selected for its high permeability to albumin. The cell loading technique was modified to ensure a high viability after injection of the cell suspension into the lumen and to minimize uneven cell distribution. Hepatocytes cultured in this bioreactor demonstrated high albumin and urea production, as well as high oxygen consumption.

For canine trials, two hollow fiber bioreactors were placed in series in the extracapillary loop of the BAL. Separate intraluminal streams were fed into each hollow fiber cartridge and discarded after a single pass. Each bioreactor was loaded with approximately 30–35 g of porcine liver cells to achieve a total liver cell mass of 60–70 g. Medium was circulated on the shell side only for the first 24 h after inoculation of hepatocytes. To determine metabolic activity of the hepatocytes, oxygen consumption and ureagenesis were monitored. For treatment, the right external jugular vein was cannulated for BAL access. The experimental animals which received D-galactosamine were divided into two groups: one which received BAL treatment, and one which received no BAL treatment. Control dogs to which no D-galactosamine was injected were included. Results indicated increased survival rates for injured canines subsequently treated with the BAL ($n = 5$) compared to injured animals which were not treated ($n = 5$) [23].

The BAL used in canine trials had a total priming volume of 800 ml; 400 ml in the extracapillary space of the bioreactors, 250 ml in the oxygenator, and the remainder in the tubing. It would be advantageous to increase the liver-specific functions as much as possible for anticipated human application. However, the amount of blood that can be drawn from a patient limits the scale-up of reactor size from that used for canine trials. Because the cell density in the device was already approaching the maximum, further increase of cell concentration in the collagen gel is not an option either. An alternative approach is to manipulate the hepatocytes such that their liver-specific activities are enhanced.

Application of hepatocyte spheroids in the BAL

Primary rat hepatocytes, when plated on certain modified surfaces, self-assemble from monolayers into multicellular aggregates, or spheroids. These freely floating spherical cell masses are typically 100 μm in diameter and six to eight cell layers thick. A distinguishing feature of spheroids is that they exhibit enhanced liver-specific functions compared to hepatocytes cultured as a monolayer [24–26]. Other distinguishing characteristics of hepatocyte spheroids include prolonged viability and a more in vivo-like ultrastructure. Spheroids remain highly viable for over a week in culture, while hepatocytes cultured as a monolayer on a collagen-coated surface or on a plastic surface typically lose more than 50% of their viability over the same time period. Spheroids appear to have smooth undulating surfaces, and individual cells are not readily distinguishable when viewed by scanning electron microscopy. Under transmission electronic microscopy, an extensive internal network of microvilli-lined bile canalicular-like channels can be seen throughout the spheroid [27]. Hepatocytes cultivated as spheroids also had higher activities of albumin biosynthesis, urea production, and cytochrome P450 enzyme activity. Hence, the desired increase in device performance may be achieved by use of spheroids in the BAL.

When plated on a surface, only 35–40% of cells form into spheroids, while the rest remain as small cell patches. If spheroids were to be prepared from monolayer culture for use in the BAL, the task of cultivating such a large number of cells would be very labor intensive. A method of preparing porcine hepatocyte spheroids in a stirred tank bioreactor was developed to alleviate this problem [28]. Freshly harvested porcine hepatocytes were inoculated at a moderate cell concentration of $3–5 \times 10^5$ cells/ml into spinner flasks containing a defined culture medium. These hepatocytes agglomerated to form small clumps within hours. Further agglomeration of these clumps resulted in spheroid formation 24 h after cell inoculation. A similar spheroid preparation technique was also developed for rat hepatocytes [29]. Approximately 80–90% of inoculated hepatocytes formed spheroids with a mean diameter of about 135 μm. Scanning and transmission electron microscopy revealed that the morphological and ultrastructural characteristics of spheroids formed in spinner flasks resembled those of spheroids prepared from monolayer culture. Advantages of preparing hepatocyte spheroids in an agitated vessel include simplicity, efficiency, reproducibility, and feasibility of producing large quantities of spheroids required for use in a BAL.

When entrapped in a collagen gel, metabolic activities of spheroids were maintained at a stable level for over 21 days [30]. Production of albumin and urea by hepatocyte spheroids entrapped in collagen gels were 2–3 times higher than those by freshly isolated single cells. P450 activity was demonstrated by metabolism of lidocaine to its main metabolite, monoethylglcinexylidide (MEGX). Phase II drug metabolism was demonstrated by glucuronidation of 4-methylumbelliferone (4-MU). This work showed that porcine hepatocyte spheroids entrapped in collagen maintained differentiated functions for an extended time period.

To explore the possibility of entrapping spheroids into the BAL, rat hepatocyte spheroids were formed in agitated vessels and loaded into the BAL [31]. At least a 4-fold improvement in albumin synthesis and P450 enzyme activity was observed compared to BAL devices containing dispersed hepatocytes entrapped in collagen gels. Subsequently, porcine hepatocytes as a mixture of spheroids and dispersed cells were used [32]. Including dispersed cells in the cell suspension increased the rate and the degree of gel contraction significantly. The spheroid-containing BAL also demonstrated approximately 2-fold increased metabolic activity on a per cell basis compared to devices entrapped with only dispersed cells.

Future directions of the hepatocyte BAL and spheroids

In the last few years, significant progress and success has been achieved in developing the BAL. The ultimate test will be to demonstrate safety and efficacy in human clinical trials. Although efforts to employ this BAL for fulminant liver failure patients have been initiated, some questions still remain. The long-term effect of the patient's exposure to porcine proteins crossing the hollow fiber membrane into the patient's blood circulation is still unknown. Furthermore, whether the 2- to 5-fold increase in measured liver-specific functions exhibited by spheroids offsets the task of preparing them needs to be evaluated. Whether or not spheroids are eventually used in a BAL, their ability to self-assemble to form structures which resemble liver cells in vivo and to maintain higher functions is attractive to many liver researchers. This process of self-assembly reverses when spheroids are transferred to highly adhesive surfaces where they then spread back into monolayers. Interestingly, the reversal of spheroidal structure and morphology coincides with the loss of cytochrome P450 activities measured in situ. With their closer resemblance to liver cells in vivo, and their ability to retain liver properties over a longer duration, spheroids, either entrapped in a BAL or cultivated in suspension culture, can be an attractive vehicle for drug metabolic research and transient gene expression studies.

Acknowledgements

This work was supported by grants from the National Science Foundation, the Whitaker Foundation, the National Institutes of Health and Regenerex Inc. The authors FJW and JRF were supported by traineeship from a Biotechnology training grant from the National Institutes of Health.

References

1. National Center for Health Statistics: Births, Marriages, Divorces and Deaths for 1990. Monthly Vital Statistics Report. Hyattsville MD: US Department of Health, Education, and Welfare, 1991;1–14.
2. Peleman RR, Gavaler JS, van Thiel DH, Esquivel C, Gordon R, Iwatsuki S, Starzl TE. Orthotopic liver transplantation for acute and subacute hepatic failure in adults. Hepatology 1987;7: 484–489.
3. Busuttil RW, Colonna JO, Hiatt JR, Brems JJ, el Khoury G, Goldstein LI, Quinones-Baldrich WJ, Abdul-Rasool IH, Ramming KP. The first 100 liver transplants at UCLA. Ann Surg 1987; 206:387–402.
4. Takahashi T, Malchesky PS, Nose Y. Artificial liver. State of the art. Dig Dis Sci 1991;36: 1327–1340.
5. Kiley JE, Welch HF, Pender JC, Welch CS. Removal of blood ammonia by hemodialysis. Proc Soc Exp Biol Med 1956;91:489–490.
6. Schechter DC, Nealon TF, Gibbon JH. A simple extracorporeal device for dreducing elevated blood ammonia levels. Surgery 1958;44:892–897.
7. Ton HY, Hughes RD, Silk DBA, Williams R. Albumin coated Amberlite XAD-7 resin for hemoperfusion in acute liver failure. Part I: Adsorption studies. Artif Organs 1979;3:20–22.
8. Opolon P, Rapin JR, Huguet C, Granger A, Delorme ML, Boschat M, Sausse A. Hepatic failure coma treated by polyacrylonitrile membrane hemodialysis. Trans Am Soc Art Int Organs 1976;22:701–710.
9. Lie TS. Treatment of acute hepatic failure by extracorporeal hemoperfusion over human and babooon liver. In: Brunner G, Schmidt FW (eds) Artificial Liver Support. Berlin: Springer-Verlag Publications, 1981;274–279.
10. Fischer M, Botterman P, Sommoggy SPS, Ernhardt W. Functional capacity of extracorporeal baboon liver perfusions. In: Brunner G, Schmidt FW (eds) Artificial Liver Support. Berlin: Springer-Verlag Publications, 1981;280–285.
11. Nose Y, Mikami J, Kasai K, Sasaki E, Agishi T, Danjo Y. An experimental artificial liver utilizing extracoporeal metabolism with sliced or granulated canine liver. Trans Am Soc Art Int Organs 1963;9:358–362.
12. Eiseman B, Norton L, Kralios NC. Hepatocyte perfusion within a centrifuge. Surg Gynecol Obstet 1976;142:21–28.
13. Matsumara K. Method and Device for Purifying Blood, US Patent No 3734851, 1973.
14. Hager JC, Carman R, Porter LE, Stoller R, Leduc EH, Galletti PM, Calabresi P. Neonatal hepatocyte culture on artificial capillaries: a model for drug metabolism and the artificial liver. ASIAO J 1983;6:26–35.
15. Shatford RA, Nyberg SL, Meier SJ, White JG, Payne WD, Hu WS, Cerra FB. Hepatocyte function in a hollow fiber bioreactor: a potential bioartificial liver. J Surg Res 1992;53:549–557.
16. Nyberg SL, Shatford RA, Peshwa MV, White JG, Cerra FB, Hu WS. Evaluation of a hepatocyte-entrapment hollow fiber bioreactor. A potential bioartificial liver. Biotechnol Bioeng 1993;41: 194–203.
17. Shatford RA, Nyberg SL, Payne WD, Hu WS, Cerra FB. A hepatocyte bioreactor as a potential bioartificial liver: demonstration of prolonged tissue specific functions. Surg Forum 1991;42: 54–56.
18. Nyberg SL, Shirabe K, Peshwa MV, Sielaff TD, Crotty PL, Mann HJ, Remmel RP, Payne WD, Hu WS, Cerra FB. Extracorporeal application of a gel-entrapment, bioartificial liver: demonstration of drug metabolism and other biochemical functions. Cell Transpl 1993;2:441–452.
19. Nyberg SL, Payne WD, Amiot B, Shirabe K, Remmel RP, Hu WS, Cerra FB. Demonstration of biochemical function by extracorporeal xenohepatocytes in an anhepatic model. Transpl Proc 1993;25:1994–1945.
20. Nyberg SL, Mann HJ, Hu MY, Payne WD, Cerra FB, Remmel RP, Hu WS. Extrahepatic

metabolism of 4-methylumbelliferone and lidocaine in an anhepatic rabbit. Drug Metab Dispos 1996;24:264—648.

21. Sielaff TD, Hu MY, Rollins MD, Bloomer JR, Amiot B, Hu W-S, Cerra FB. An anesthetized model of lethal canine galactosamine fulminant hepatic failure. Hepatology 1995;21:796—804.
22. Sielaff TD, Hu MY, Rao S, Groehler K, Olson D, Mann HJ, Remmel RP, Shatford RA, Amiot B, Hu WS, Cerra FB. A technique for porcine hepatocyte harvest and description of differentiated metabolic functions in static culture. Transplantation 1995;59:1459—1463.
23. Sielaff TD, Hu MY, Amiot B, Rollins MD, Rao S, McGuire B, Bloomer JR, Hu W-S, Cerra FB. Gel-entrapment bioartificial liver therapy in galactosamine hepatitis. J Surg Res 1995;59: 179—184.
24. Koide N, Shinji T, Tanabe T, Asano K, Kawaguchi M, Sakaguchi K, Koide Y, Mori M, Tsuji T. Continued high albumin production by multicellular spheroids of adult rat hepatocytes formed in the presence of liver-derived proteoglycans. Biochem Biophys Res Commun 1989;161:385—391.
25. Tong JZ, De Lagusie P, Furlan V, Cresteil T, Bernard O, Alvarez F. Long-term culture of adult rat hepatocyte spheroids. Exp Cell Res 1992;200:326—332.
26. Peshwa MV, Wu FJ, Follstad BD, Cerra FB, Hu WS. Kinetics of hepatocyte spheroid formation. Biotechnol Progr 1994;10:460—466.
27. Peshwa MV, Wu FJ, Sharp HL, Cerra FB, Hu W-S. Mechanistics of formation and ultrastructural evaluation of hepatocyte spheroids. In Vitro Cell Dev Biol 1996;32A:197—203.
28. Lazar A, Peshwa MV, Wu FJ, Chi CM, Cerra FB, Hu WS. Formation of porcine hepatocyte spheroids for use in a bioartificial liver. Cell Transpl 1995;4:259—268.
29. Wu FJ, Friend JR, Hsiao C-C, Zilliox MJ, Ko W-J, Cerra FB, Hu W-S. Efficient assembly or primary rat hepatocyte spheroids for tissue engineering applications. Biotechnol Bioeng 1996;50: 404—415.
30. Lazar A, Mann HJ, Remmel RP, Shatford RA, Cerra FB, Hu W-S. Extended liver-specific functions of porcine hepatocyte spheroids entrapped in collagen gel. In Vitro Cell Dev Biol An 1995;31:340—346.
31. Wu FJ, Peshwa MV, Cerra FB, Hu W-S. Entrapment of hepatocyte spheroids in a hollow fiber bioreactor as a potential bioartificial liver. Tis Eng 1995;1:29—40.
32. Wu FJ, Friend JR, Mann HJ, Remmel RP, Cerra FB, Hu W-S. Hollow fiber bioartificial liver utilizing collagen entrapped porcine hepatocyte spheroids. Biotechnol Bioeng 1996;52:34—44.

Tissue Engineering for Therapeutic Use 1.
Y. Ikada and Y. Yamaoka, editors.

Application of human recombinant morphogenetic protein (rhBMP-2) in the reconstruction of mandibulectomy and maxillectomy defects

Philip J. Boyne
Department of Surgery, Division of Oral and Maxillofacial Surgery, Loma Linda University Medical Center, California, USA

Abstract. Molecular cloning of the family of bone morphogenetic proteins (BMP) has led to the possibility of the therapeutic utilization of these related substances in the reconstruction of large maxillofacial defects. Of the members of the TGF-β superfamily of proteins available, recombinant human bone morphogenetic protein (rhBMP-2) was selected for our investigation. This group of inductor proteins has been shown to increase the expression of markers of osteoblastic phenotypes in certain cell lines [1]. Also the BMP-2 had been shown to have an effect on earlier osteoprogenitor cells rather than merely on the more differentiated cell lines. We wished to evaluate the effect of this inductor substance in larger osseous defects rather than in those recipient sites previously evaluated in laboratory studies. The purpose of the series of investigations being discussed here was to determine the effect of rhBMP-2 on the regeneration of hemimandibulectomy defects of mature subhuman primates, and the effect of function on the maturation and remodeling of the newly formed bone. Further we wished to study the effect of rhBMP-2 on simulated maxillary bony clefts in growing subhuman primates.

The application of human recombinant BMP to these defects involves the development of technical and surgical procedures which will allow the inductor agent to interact with the appropriate host osteoprogenitor cells. Development of the optimally effective surgical model to allow the inductor material to accomplish this interaction is at times very difficult. In our studies we approached this problem by the application of two models (the hemimandibulectomy defect and the cleft palate osteotomy). In this manner, we hoped to demonstrate that the appropriate surgical isolation of the BMP inductor in a suitable carrier with opportunity to interact with available host osteoprogenitor cell populations and the postoperative maintenance of surgical "space" is essential to the optimal application of bone induction.

Keywords: induced bone, maintenance, mandibulectomy, rhBMP-2, space, surgical isolation.

Introduction

Approximately three decades ago, Dr Marshall Urist's work indicated that there was a very real possibility of induction of bone in heterotopic sites by implantation of certain protein extracts [1–3]. The protein moiety used for such induction was termed "bone morphogenetic protein" or BMP. The material was further characterized, and assayed amounts of BMP were found to induce heterotopic

Address for correspondence: Philip J. Boyne DMD MS DSc(Hon), Loma Linda University, School of Dentistry and Medicine, Department of Oral & Maxillofacial Surgery, 11092 University Court, Room 3306, Loma Linda, CA 92350, USA. Tel.: +1-909-824-4671. Fax: +1-909-478-4285.

bone formation in the rat [3]. There appeared to be a cartilaginous phase of the development of the bone ossicle at the ectopic sites. Additionally, bone formation in heterotopic sites was somewhat bizarre and disorganized. Questions regarding the effectiveness of the use of such an inductive material in large defects, and whether such possibly resultant reparative bone could withstand the effects of normal function and appropriately remodel to form an anatomic bone structure consistent with the pre-existing bone remained of paramount clinical interest.

The existence of a pool of osteoprogenitor cells available for recruitment and for phenotypic change by morphogenetic induction is essential to the initial success and long-term effectiveness of the system. Furthermore, the pool of cells must be maintained and available for an appropriate length of time. Surgically created injury and healing events such as primary and secondary hemorrhage, edema, movement of the host bone fragments and other types of local as well as systemic factors could adversely affect the action of the bone induction being evaluated. For this reason we developed surgical systems to attempt to provide the optimal environment for the evaluation of the BMP. These models included from the surgical standpoint:

1. The development of a critical size defect (one not subject to spontaneous regeneration without the aid of a rather massive osteogenetic bone graft).
2. The maintenance by orthopaedic fixation devices of the critical size defect throughout the healing phase without constrictive change in anatomic configuration of the defect and without orthopaedic motion at the surgical site.
3. The maintenance of a soft tissue "capsule" for the maintenance of the BMP and the carrier within the defect so that the effect of the BMP would not be dissipated into the surrounding soft tissues.
4. The avoidance of hemorrhage and formation of large hematomas and edema in the area which would further alter the effect of the rhBMP-2.
5. Appropriate postoperative "stressing" of the surgical site by functional stimuli normally received by the mandible and/or maxilla.

Methods

Hemimandibulectomy defects in subhuman primates to evaluate BMP dose effectiveness

In a group of seven animals bilateral discontinuity, full thickness defects were made in excess of 2.2 cm in length. These were created by use of high speed orthopaedic cutting instruments using copious irrigation in the mandible commensurate with the usual mandibular resection undertaken in human oncologic surgical procedures. The male *Macaca fascicularis* monkeys were between 8 and 11 years of age and in excellent health.

Following the resections, a titanium orthopaedic mesh (Ti-Mesh, Safomor Danek Corp, Memphis, Tenn, USA) was inserted to maintain the space in the correct distracted position (Fig. 1). This model defect has been shown by previous

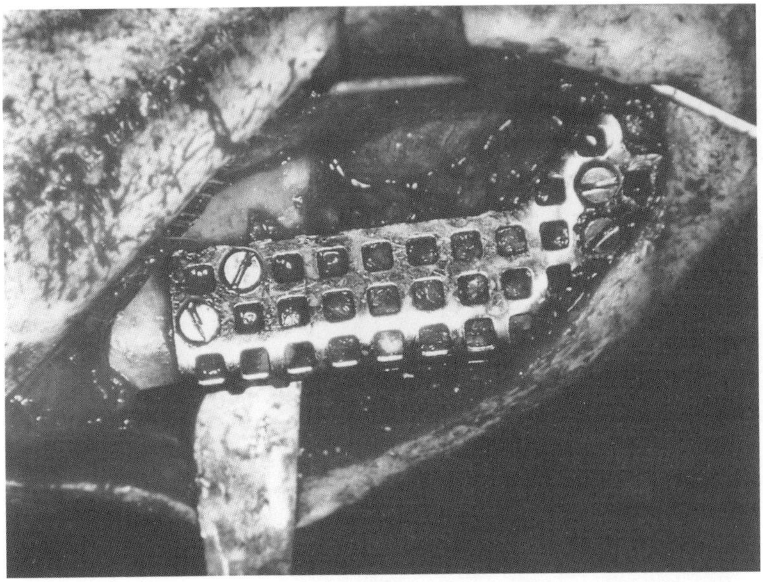

Fig. 1. A surgical hemimandibulectomy defect in a mature *Macaca fascicularis* monkey. The defect is in excess of 2.2 cm in length and is stabilized and maintained by a titanium orthopaedic mesh.

work to be incapable of complete regeneration or even partial regeneration without the use of a very extensive autogenous bone graft [4]. Following the resection, the animals received rhBMP-2 (recombinant human bone morphogenetic protein) in a collagen carrier sponge (Helistat, Collatec Inc., Plainsboro, New Jersey, USA) (Fig. 2). The tissues were closed in the layers and the animals were allowed to function on a soft diet for 4 months.

Previous to mandibular resection surgery the animals had received extraction of the teeth bilaterally including cuspid, premolars, and first and second molar teeth. The third molar tooth and the incisors were allowed to remain in place to establish appropriate occlusion. The soft tissue flap was sutured creating a complete mucoperiosteal closure which was allowed to heal for 4 weeks before the mandibular resection was undertaken. This procedure produced the superiorly closed periosteal portion of the soft tissue envelope, enabling the BMP and the carrier to be placed within a space entirely enclosed by soft tissue or the osteotomized ends of the bone fragments and the titanium orthopaedic mesh itself inferiorly. The development of this "dead space" was felt to be essential to the creation of an appropriate receptive model.

Of this group of animals three had been designated to evaluate a "high" vs. "low" rhBMP-2 dose to the defect. One side received a dose of 0.8 mg of BMP per cc of solution and the contralateral defect received 0.2 mg per cc of the solution. All animals were appropriately labeled with tetracycline given intravitally (intramuscularly) to mark the reparative bone [4]. The animals were sacrificed 5 months postsurgically.

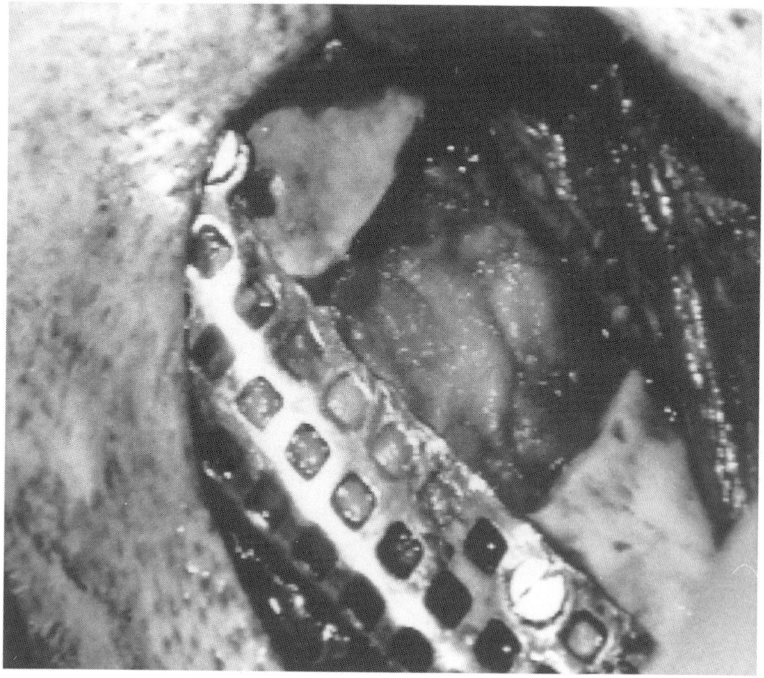

Fig. 2. A collagen sponge (helistat) treated with rhBMP-2 is placed in the defect on the right side. A similar defect on the left side hemimandibulectomy received the same treatment with a lower rhBMP-2 dose.

Long-term comparison of rhBMP-2 implanted vs. PMCB grafted hemimandibulectomy defects

Of the *Macaca fascicularis* animals four were maintain on a long-term experiment with a protocol modification which involved having one hemimaxillectomy defect in each animal receive 0.75 mg of rhBMP-2. The contralateral side with a comparable defect received autogenous particulate marrow and cancellous graft (PMCB).

This group of animals was operated to observe the comparison between particulate marrow cancellous bone grafted hemimandibulectomy defects and the BMP-implanted defects, and also to observe the effect of root form implants placed in these areas (Fig. 3). Implants were placed at the end of 4 months on both sides and the animals were appropriately sacrificed after 6 months of functioning with the root form supported prosthesis (placed 5 months after implant insertion). Thus, the effect of the animal functioning with a normal diet on root form implants for 6 months and the effect on the maturation and remodeling of the BMP regenerated defect vs. the autogenous bone regenerated defect was observed.

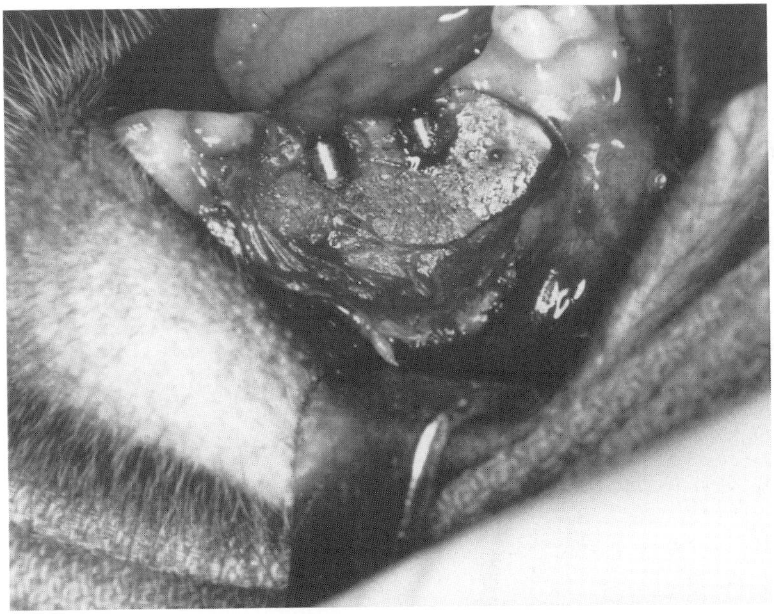

Fig. 3. Implants placed in the rhBMP-2 treated side will show excellent remodeling of the regenerated area — at the 4-month level. The 2.5-cm hemimandibulectomy defect has completely regenerated showing excellent continued bone formation.

RhBMP-2 in maxillary cleftal defects

In order to evaluate the effect of rhBMP-2 in the maxilla, defects were produced which simulated a bilateral cleft palate in six *Macaca mulatta* monkeys 1½ years of age. This age group is chronologically comparable to a 5- to 6-year-old human child [5]. The surgical "cleft" simulating defects were made by osteotomy from the piriform aperture of the maxilla to the alveolar ridge following the removal of the incisor and permanent deciduous anterior teeth (Fig. 4). An osteotomy was also made across the palate to complete the bilateral cleft-like defect. The oral mucosa was then closed to the nasal mucosa to deliberately produce oral nasal fistulas.

After the 4 months of healing the animals were found to have residual fistulas which were still patent (Fig. 5). The defects of the animal's maxilla resembled that of a bilateral cleft palate child. At this point the mucoperiosteal flaps were reopened by incision. The bone was found not to have regenerated and there was no evidence of bony healing in the surgically created cleftal defect walls. The animals were then implanted on one side with the BMP with a dose of 0.75 mg/cc. On the contralateral side, autogenous particulate marrow and cancellous bone was grafted. The animals were allowed to function for an additional 3 months at which time, after receiving appropriate labels of intravital tetracycline (15 mg per kg of body weight), sacrifice of the animals was accomplished and the tissues were evaluated histomorphometrically.

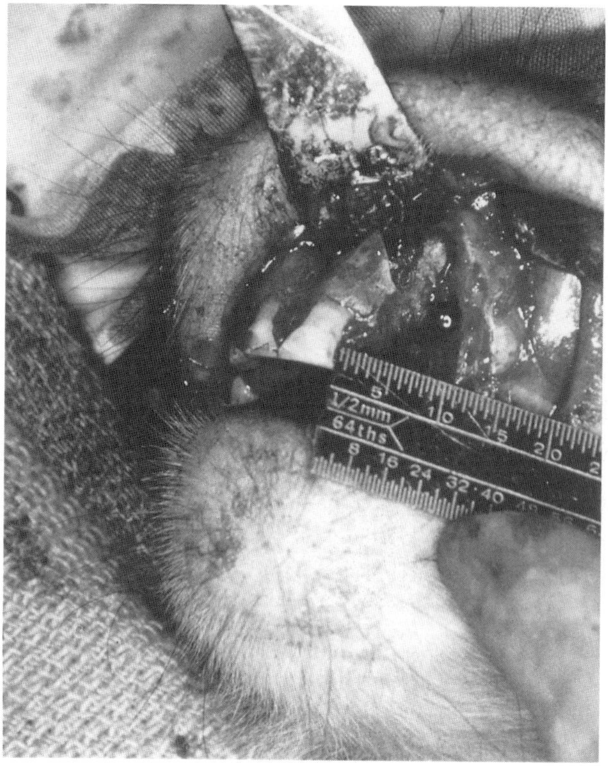

Fig. 4. A surgically created maxillary cleft defect showing an osteotomy 8 mm in width extending from the pyriform aperture to the alveolar ridge.

Results

Mandibular defects — effect of BMP-2 dose

Both of the defects (the high vs. low BMP-dose defects) were found to have regenerated in an excellent manner with a slight difference noted between the high and the low dose area [4].

The defects receiving 0.8 mg of rhBMP-2 per cc revealed clinically a slight increase in the formation of bone matrix at 5 months postoperative compared with the low-dose size, although the difference was not significant (Fig. 6). Histomorphometric analysis of the amount of area of calcified bone matrix and marrow vascular spaces in the rhBMP-2-implanted defect indicated no significant difference between the high- and low-dosed defects [4].

Mandibular defects — effect of function

The effect of titanium root form implants on the regenerated bone was one of

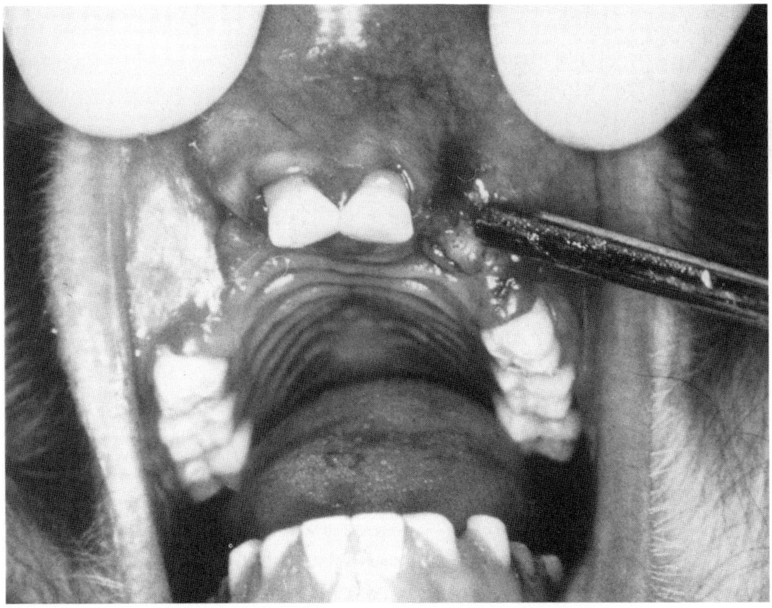

Fig. 5. After 4 months of healing, the osteotomy site is shown with fistulas bilaterally. The defect is patent and no osseous regeneration has taken place.

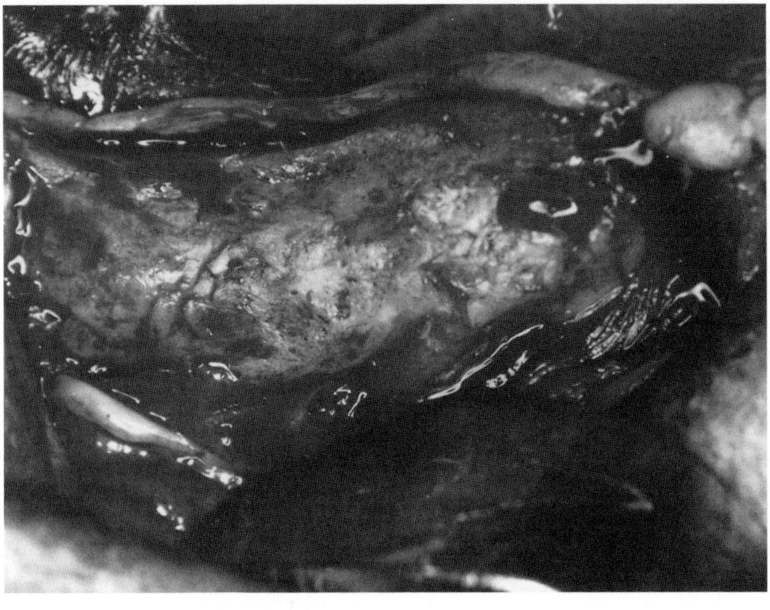

Fig. 6. Complete osseous regeneration of the mandibulectomy defect is apparent after surgical exposure intraorally at 4 months.

enhanced cancellous bone production resulting in a increase in thickness of the trabecular pattern, and an increase thickness of the cortical bone at the alveolar crest (Fig. 7). Thus, the general impression was one of a favorable bone remodeling response in the BMP area. The enhanced remodeling appeared to be influenced by the stimulating effect of the root form implants.

Data on the actual amount of increase in thickness of the trabecular pattern of the cancellous bone produced by the function of the root form implants are currently being evaluated. However, review of the specimens histologically revealed no large marrow-vascular (MV) spaces as were seen in the earlier animals who received the BMP but were not stimulated by root form implant function.

Maxilla cleftal defects

The portion of the simulated cleft defect extending from the piriform aperture of the nose to the crest of the alveolar ridge showed a marked increase in cortical bone surface remodeling over the osseous repaired defect of the palatal side. The autografted side showed the same type of restoration. Statistically there was no difference in the regeneration of the lateral maxillary bony defects between the two graft methods used, e.g., PMCB vs. rhBMP-2 (Fig. 8).

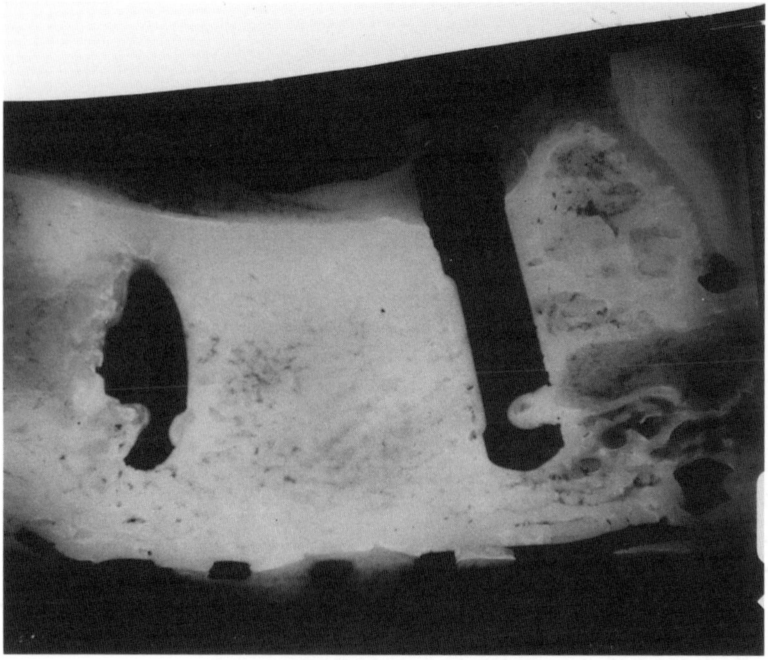

Fig. 7. Specimen showing implants in place in the regenerated mandibule in function for 6 months. There has been a consolidation and an increase in density of the cancellous bone adjacent to implants. Tetracycline-induced fluorescence is demonstrating bone remodeling (magnification × 4).

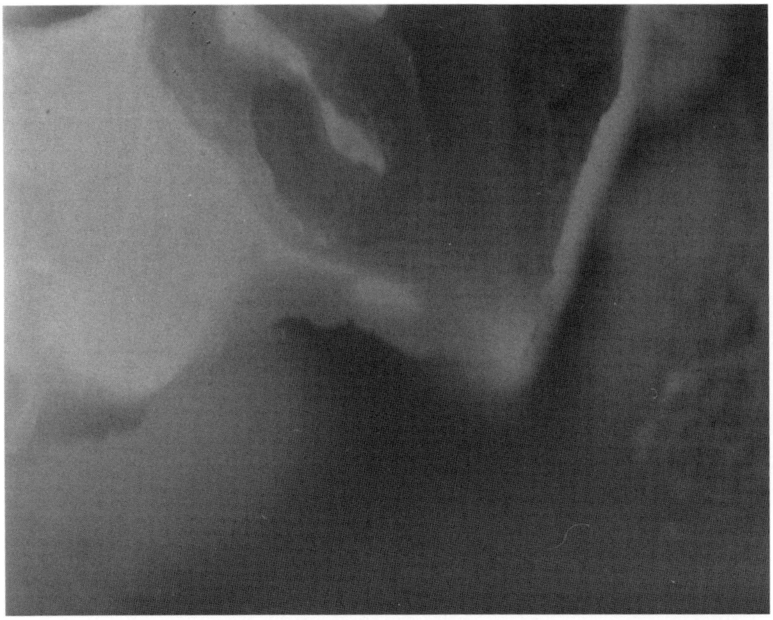

Fig. 8. A specimen demonstrates the complete healing of the cleft on the right side of the figure (rhBMP-2) at 3 months while the PMCB-grafted side (left) is still not completely reconstructed with new bone. While the PMCB graft will eventually regenerate the defect, the rate of reconstruction is superior on the rhBMP-2 side (magnification × 4).

The regeneration on the palatal aspect of the defect was in all cases more advanced on the BMP-implanted side. Thus the rhBMP-2 was at least equal in effecting osseous reconstruction of the simulated cleft defects at the 3-month postoperative level compared with the presently accepted method of cleft grafting by autogenous particulate cancellous bone from the iliac crest [6].

Histomorphometric evaluation in both the mandibular study and the cleft palate evaluation was made on a semiautomatic computer with an image analysis system (SNI Unicomp, Atlanta, Georgia, USA). The system consisted of a microscopic-attached computer equipped with a mounted video camera monitor. The system image was projected through the microscope onto a video screen and was traced on a digitalizer pad. Conversion was made from the image to a known area of measurement.

For the mandibular resection study two areas were evaluated:

1) the central position of the reconstructed defect composing total areas of calcified bone matrix in association with marrow vascular spaces, and
2) the thickness of the reformed cortical bone surface.

In the maxillary cleftal defects, two anatomic areas were also compared:

1) the lateral surface of the maxilla from the piriform aperture to the crest of the alveolar ridge (Fig. 8), and

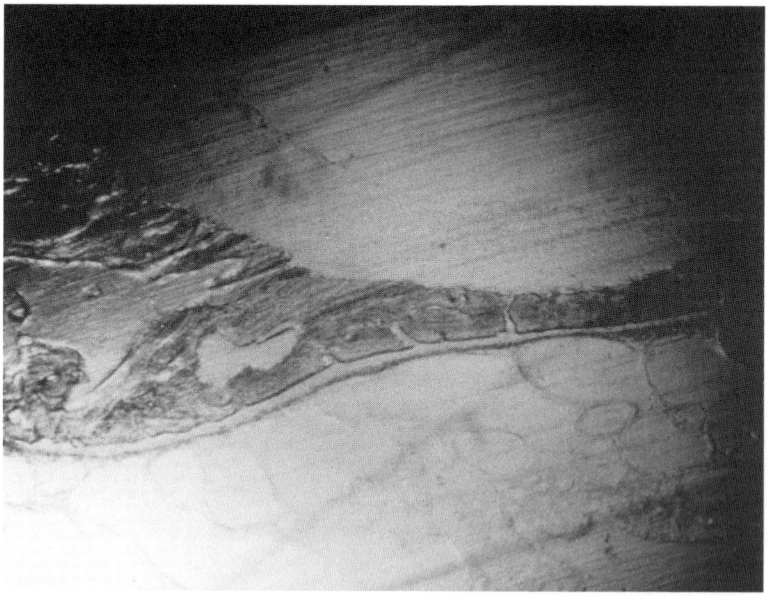

Fig. 9. View of palatal aspect of the cleft defect on the rhBMP-2-implanted side. There has been a complete regeneration of the defect and a reformation of bone cortex.

2) the palatal portion of the crest from alveolus to the palatal suture (Fig. 9) [6]. In the initial mandibular study, representative sections for the regenerated defects were taken for calculation of trabecular bone along the marrow vascular spaces comparing the high- with the low-dose side.

Conclusion

In these animal studies carried out in both mature and immature *Macaca mulatta* and *Macaca fascicularis* monkeys, rhBMP-2 was effectively used to produce osseous regeneration of critical sized maxillary and mandibular defects.

The bone of the regenerated areas underwent remodeling and increased the overall bone density of both the cortical surfaces of the jaws and the trabecular pattern.

The reconstructed area also responded appropriately to function imposed on root form implants placed in the reformed bone for a period of 6 months.

Design of the surgical defects and maintenance of the surgical space is critical to the successful application of rhBMP-2 induction in reconstruction.

References

1. Wozney J. The Bone morphogenetic protein family and osteogenesis. Molec Reprod Devel 1992; 32:160–167.

2. Urist MR. Bone formation by autoinduction. Science 1965;150:893–899.
3. Urist MR, Mikulski A, Lutz A. Solubilized and insolubilized bone morphogenetic protein. Proc Natl Acad Sci USA 1979;76:1828–1832.
4. Boyne PJ. Animal studies of the application of rhBMP-2 in maxillofacial reconstruction. Bone 1996;(July) 19(Suppl 1):825–925.
5. Boyne PJ. Use of marrow-cancellous bone grafts in maxillary alveolar and palatal clefts. J Dent Res 1974;53;4:821–824.
6. Boyne PJ, Nath R and Nakamura A. Human recombinant BMP-2 in osseous reconstruction of simulated cleft palate defects. Br J Oral Fac Maxillofasc Surg 1998;36:84–90.

Tissue Engineering for Therapeutic Use 1.
Y. Ikada and Y. Yamaoka, editors.

Incorporation of physical factors into tissue engineering for therapeutic use

Takashi Ushida[1,2] and Tetsuya Tateishi[1,2]
[1]*3D Tissue Engineering Group, National Institute for Advanced Interdisciplinary Research, Tsukuba, Ibaraki; and* [2]*Institute of Medical Research, Graduate School of Medicine, University of Tsukuba, Tsukuba, Ibaraki, Japan*

Keywords: G-protein, hydrostatic, integrin, poking, SA channel, shear.

Introduction

It is well-known that living tissues are under the influence of several kinds of physical factors, and regulated by them. For example, bone and cartilage are loaded with weight and exercise, and bone remodeling and cartilage functions are regulated at the cellular level. In the same way, because a blood vessel is under blood flow and pulsatile blood pressure, endothelial cells are under shear stress, and endothelial cells and smooth muscle cells are periodically stretched. These factors are known to regulate cellular functions at the molecular revel.

On the other hand, it is still unknown how physical factors such as shear stress are transduced as a signal into the cytoplasm, although there is much research on the relationship between input of physical factors and cellular responses as output. Therefore, the question is to elucidate at cellular and molecular levels the mechanism of how such a signal transduction is realized.

Effects of physical factors on tissues and cultured cells

A number of research studies have been devoted to finding the relations between input of physical factors and cellular responses as output. Those studies have been carried out by using both experimental animals and cultured cells. Among them, representative research methods are as follows: mechanical deformation (compression or stretching) loading on bone and osteoblast, hydrostatic pressure loading on chondrocyte, and shear stress loading on endothelial cell.

Effects of hydrostatic pressure on cartilage and chondrocyte [1–8]

It is known that chondrocytes respond to loading, changing the composition, mechanical properties and structure of articular cartilage. Loading is also known

Address for correspondence: Takashi Ushida, 3D Tissue Engineering Group, National Institute for Advanced Interdisciplinary Research, 1-1-4 Higashi, Tsukuba, Ibaraki 305, Japan.

to change various physical factors, such as hydrostatic pressure, osmotic pressure, extracellular ionic composition, gas concentration, and streaming potential. It has been suggested that among them hydrostatic pressure is mainly responsible for the regulation of extracellular matrix (ECM) synthesis. Substantial effects of hydrostatic pressure are, however, still obscure. Experiments in vitro have frequently used slices of articular cartilage for the evaluation of the effects of mechanical forces on cartilage metabolism. The transduction mechanisms for pressure loading in chondrocytes are unclear, but effects of hydrostatic pressure have been examined in vitro with cartilage slices and with monolayers of isolated chondrocytes. The effects of hydrostatic pressure on sulfate and proline incorporation by slices of bovine articular cartilage depend on the magnitude and duration of pressure. Those investigations concluded that cell/matrix interactions influence the effects of cyclic hydrostatic pressure on cellular function. Studies with isolated chondrocytes also show biphasic effects on proteoglycan synthesis and aggrecan mRNA expression so that high static pressure is inhibitory, while cyclic pressure at 0.5 Hz is stimulatory and cyclic pressure at 0.017 Hz is inhibitory, each at the transcriptional and translational levels.

Effects of shear stress on endothelial cell [9—18]

Endothelial cells are situated at the inner surface of blood vessels, forming a monolayer which is in direct contact with blood flow. It is known that the endothelial cell is not only an antithrombogenetic barrier, but also plays critical roles in vascular growth, and remodeling and regulating vascular tonus. Additionally, these functions are also regulated by shear stress. It is reported that shear stress stimulates proliferation and migration of endothelial cells, and changes their morphology and orientation. Shear stress is also known to influence gene expression of the following substances: PDGF, HB-EGF, bFGF, NOS, ICAM-1, VCAM-1, tPA, c-Fos, c-Myc, and c-Jun.

Signal transduction mechanism of physical factors

Although much research has been done to elucidate the relationship between various input of physical factors and cellular functions, it is still unknown what the molecular mechanism is in the signal transduction of mechanical stress into the cytoplasm (Fig. 1). For the moment, three putative mechanism are shown.

ECM-integrin-cytoskeleton complex [19—23]

Cells adhere to the extracellular matrix at adhesion plaques. At the adhesion plaques, the integrin interacts with the ECM and cytoskeleton. Therefore, it is reasonable to think that the adhesion plaque is an anchor point and transduces mechanical stress to cytoskeletons such as the actin filament (Fig. 2). The possibility that ECM receptors, such as members of the integrin family, mediate

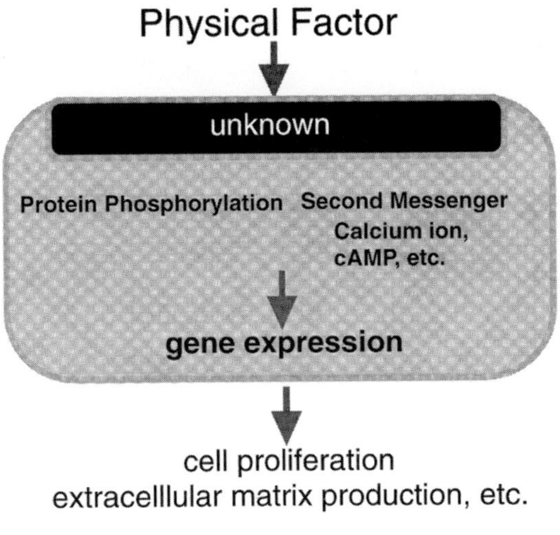

Fig. 1. Schematical pathway of signal transduction from input of physical factors to cellular responses.

mechanotransduction is supported by the finding that stretching flexible bottoms to which cells adhere alters cytoskeleton organization and induces biochemical changes in adherent cells.

Stretch-activated ion channel [24–31]

The patch-clump method is frequently used by researchers in the field of physiology for measuring ion channel currents, and they also observe channel currents

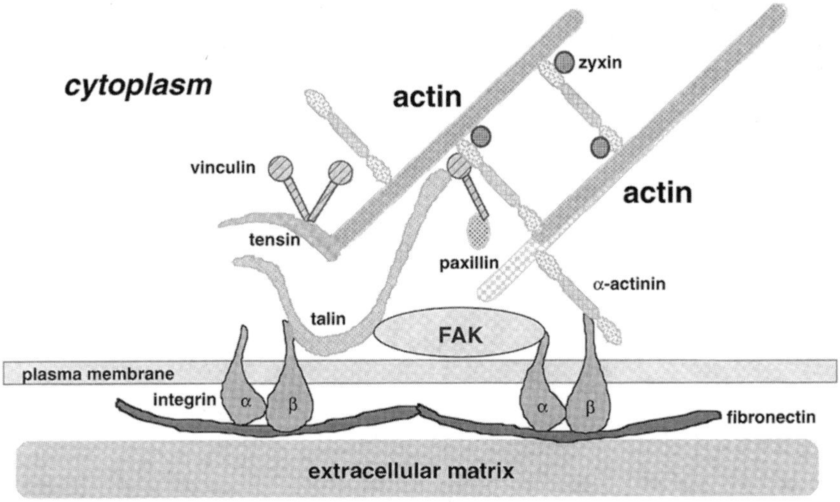

Fig. 2. Molecular interactions and complex at cell-ECM adhesion plaque. FAK: focal adhesion kinase. (Modified from Otey CA, J Biol Chem 1993;268:21193–21197.)

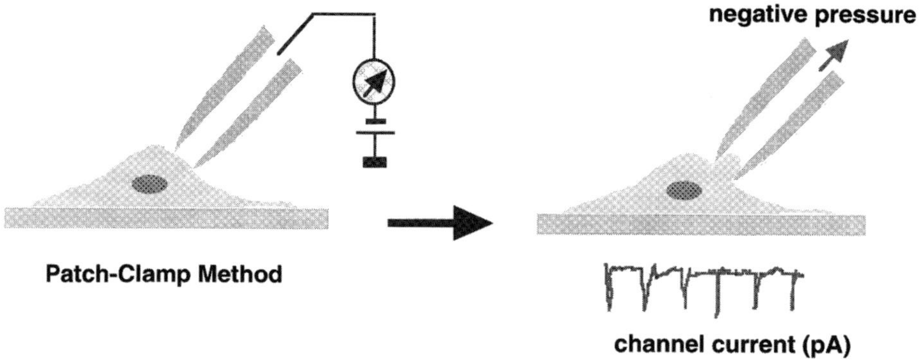

Fig. 3. Stretch-activated channel measurement using the patch-clamp method.

by the sucking of cellular membrane (Fig. 3). They have tried to measure channel currents of a whole cell with the whole cell clamp mode. However, it is difficult to keep the sealing under a positive pressure. Therefore, the currents of a single ion channel are measured in cell attached mode. Those channels detected by the method are mainly nonselective cation channels, the conductance of which is about 20—40 pS. The currents of those channels are known to be blocked with gadolinium ions. They then presume the existence of a stretch-activated ion channel on the plasma membrane.

GTP-binding-protein-mediated signal transduction [32—38]

Several kinds of transmembranous signal transduction pathways initiated by ligand-receptor interactions are known. One of the major pathways is that leading to the release of calcium ions from intracellular calcium stores (Fig. 4). These

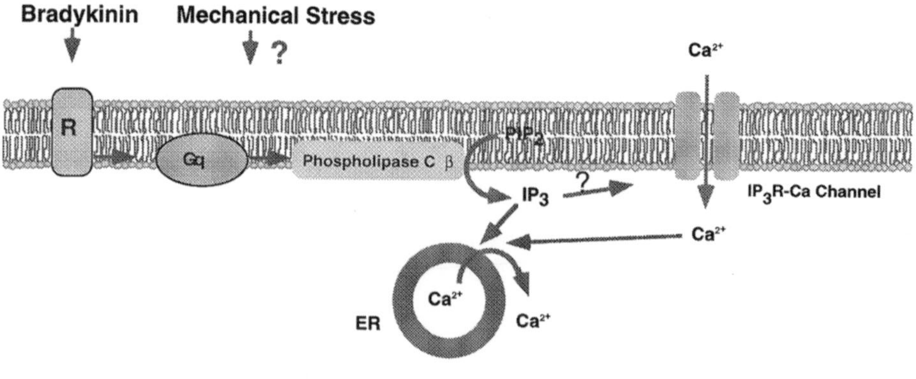

Fig. 4. Molecular mechanism of ligand-receptor interaction leading to calcium ion release through GTP-binding protein and phospholipase C activation. R: receptor; G: GTP-binding protein; PIP2: phosphatidylinositol 4,5-bisphosphate; IP3: inositol 1,4,5-triphosphate; ER: endoplasmic reticulum.

kinds of receptors leading to calcium ion release are coupled with guanosine triphosphate (GTP)-binding proteins. These GTP-binding proteins are heterotrimeric proteins composed of α, β and γ subunits. The GTP-binding proteins are activated by morphological changes of receptors with ligand-receptor interactions. The activated α subunit mainly activates phospholipase C-β (PLCβ). This phosphoinositide-specific phospholipase C generates two intracellular messengers, diacylglycerol (DG) and inositol 1,4,5-triphosphate (IP_3). IP_3 is known to release calcium ions from endoplasmic reticulums (ER), intracellular calcium stores, by interacting with IP_3-receptors on the ER, which form calcium channels. Calcium channels formed by IP_3-receptors on the plasma membrane are also presumed. The release of calcium ions transiently increases the concentrations of cytosolic calcium ions (Ca transient), leading to the activation of signal transduction cascades at the lower stream, resulting in the regulation of gene expression.

The concentrations of cytosolic calcium ions are usually quantified by using a double excitation wavelength method with a fluorescent calcium indicator, Fura-2. This method is based on the characteristic of this indicator that the fluorescence intensity at 340 nm of excitation wavelength becomes stronger according to the increase of concentration of calcium ions, while the fluorescence intensity

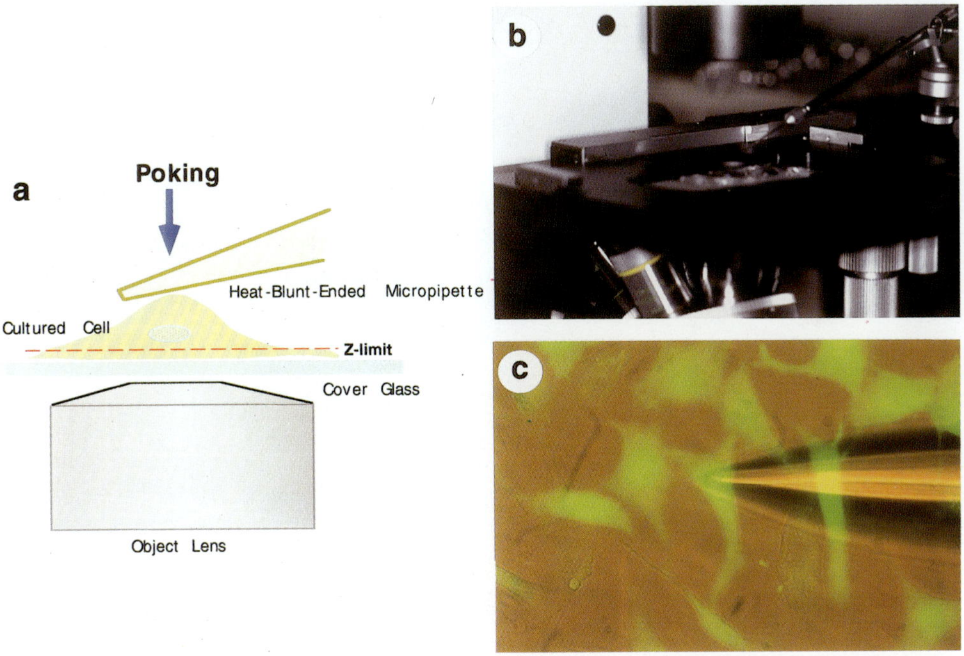

Fig. 5. Schematic representation for poking a single cell by using a micropipette. **A:** Cross-section image of a cell poking system. **B:** Micropipette with a micromanipulator set to inverted fluorescence microscopy. **C:** Transparent image of a micropipette and fluorescent image of cultured cells stained with Fluo-3.

at 360 nm stays constant. Then, if the ratio of the fluorescence intensity at 340 nm to that at 360 nm is calculated, the ratio has a good linear relationship with the concentration of calcium ions.

The time-course changes of cytosolic calcium ion concentration could be quantified and visualized by using an intensified-charged-coupled-device (I-CCD) camera for inputting fluorescent images of cells and an image processing system with a large capacity of RAM. In order to observe calcium ion concentration changes caused by a physical factor within a single cell, it is necessary to load a single cell with that factor. One solution under this restriction is to poke a single cell with a micropipette set to a micromanipulator on the stage of an inverted fluorescence microscope (Fig. 5). The propagation of Ca transients from a poking point within a cell could be detected immediately after poking the point (Figs. 6 and 7). The propagation of Ca transients to neighboring cells could be also detected.

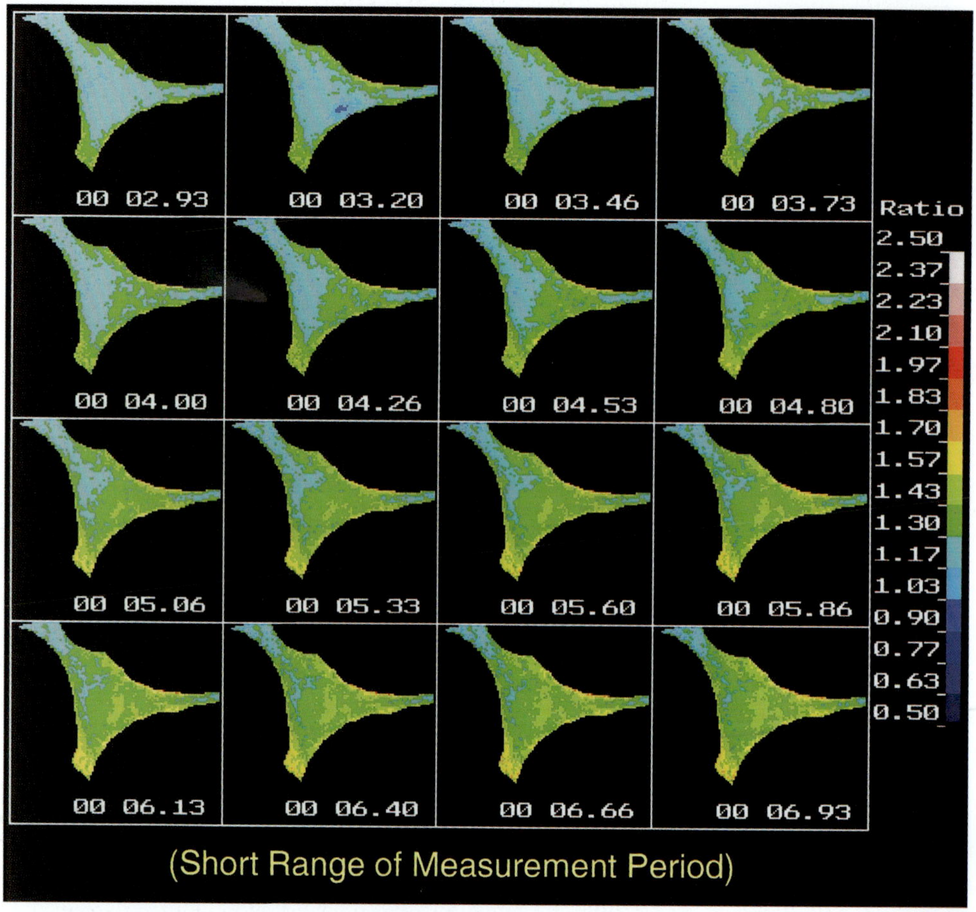

Fig. 6. Time course changes of intracellular free Ca^{2+} concentrations by poking a bovine endothelial cell.

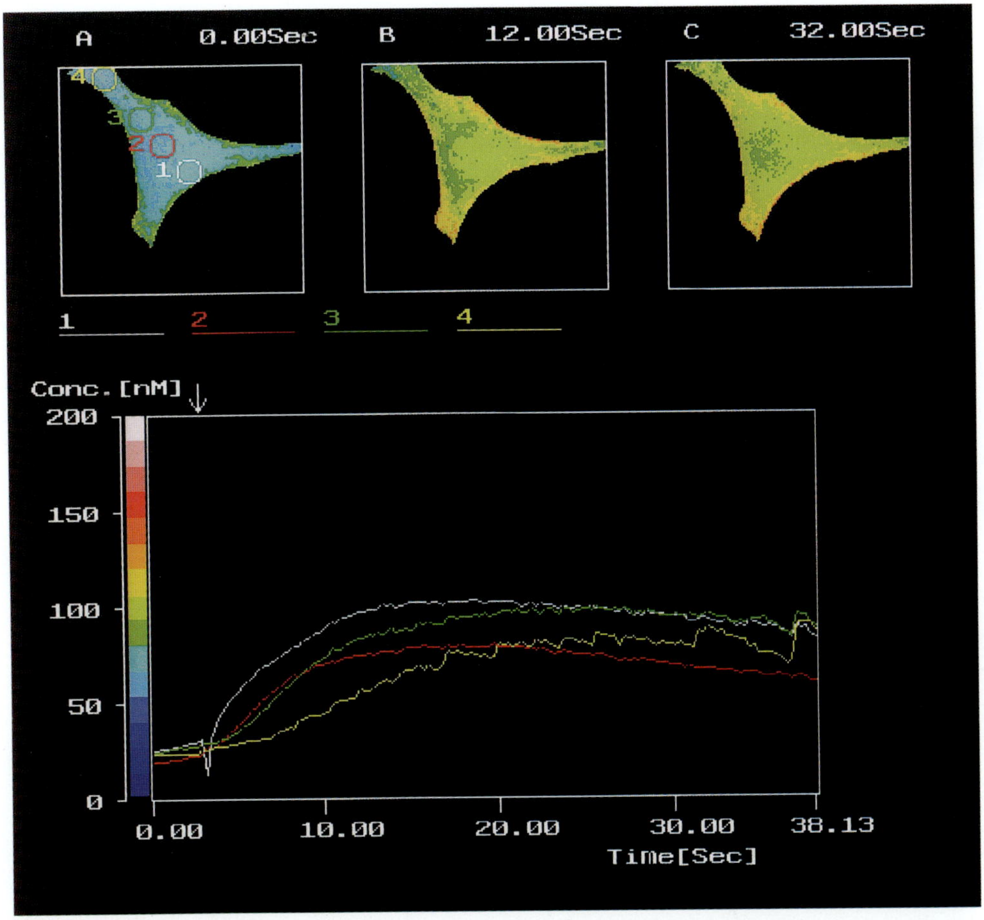

Fig. 7. Time course changes of intracellular free Ca^{2+} concentrations by poking a bovine endothelial cell.

When cultured cells are incubated for 30 min with 1.0 μM of U-73122, an inhibitor of phospholipase C (PLC), the addition of bradykinin (polypeptide, a ligand to a specific receptor that is coupled with a GTP-binding protein) into the medium does not cause Ca transients. Under the same conditions for which the activity of PLC is estimated to be inhibited, Ca transient is also not observed by poking the cells. Then calcium ions in the culture medium are chelated by replacing it with PBS-5 mM EGTA. Under this condition, Ca transient is observed neither by adding bradykinin nor by applying mechanical stress. The results show that it is possible that the activation of PLC and calcium influx would be involved in Ca transient by loading cultured cells with mechanical stress.

Incorporation of physical factors into tissue engineering for therapeutic use

As is well-recognized, two kinds of factors are indispensable for reconstructing living tissues in vitro. One factor is the materials, e.g., biomaterials like collagen, laminin, biodegradable material, and nonbiodegradable material. Because cell adhesion is known to regulate cellular functions through the formation of an ECM-integrin complex and subsequent FAK (focal adhesion kinase) phosphorylation, materials for cell adhesion substrate are thought to be essential.

The other factor is chemicals such as adequate growth factors, cytokines and other bioactive molecules. The presence of adequate bioactive molecules in an appropriate space at the right time could be fundamental for reconstructing living tissues in vitro.

Further, a possible third factor is whether or not physical factors are necessary in tissue engineering. As described above, living tissues are under the influence of various kinds of physical factors and regulated by them in vivo, and many kinds of cultured cells respond to physical factors and change their cellular functions in vitro. According to this context, we think that physical factors have a supplementary role in constructing specific kinds of living tissues in vitro, considering the experimental results, and that the incorporation of physical factors could be indispensable for reconstructing these kinds of tissues in vitro.

However, it is still unknown how physical factors such as mechanical stress are transduced into the cytoplasm. It is essential to elucidate a signal transduction mechanism of physical factors in order to incorporate them totally into tissue engineering for therapeutic use.

References

1. Parkkinen JJ, Lammi MJ, Inkinen R, Jortikka M, Tammi M, Virtanen I, Helminen HJ. Influence of short-term hydrostatic pressure on organization of stress fibers in cultured chondrocytes. J Orthop Res 1995;13:495–502.
2. Parkkinen JJ, Lammi MJ, Pelttari A, Helminen HJ, Tammi M, Virtanen I. Altered Golgi apparatus in hydrostatically loaded articular cartilage chondrocytes. Ann Rheum Dis 1993;52: 192–198.
3. Parkkinen JJ, Ikonen J, Lammi MJ, Laakkonen J, Tammi M, Helminen HJ. Effects of cyclic hydrostatic pressure on proteoglycan synthesis in cultured chondrocytes and articular cartilage explants. Arch Biochem Biophys 1993;300:458–465.
4. Lammi MJ, Inkinen R, Parkkinen JJ, Hakkinen T, Jortikka M, Nelimarkka LO, Jarvelainen HT, Tammi M. Expression of reduced amounts of structurally altered aggrecan in articular cartilage chondrocytes exposed to high hydrostatic pressure. Biochem J 1994;304:723–730.
5. Hall AC, Urban JP, Gehl KA. The effects of hydrostatic pressure on matrix synthesis in articular cartilage. J Orthop Res 1991;9:1–10.
6. Kim YJ, Sah RL, Grodzinsky AJ, Plaas AH, Sandy JD. Mechanical regulation of cartilage biosynthetic behavior: physical stimuli. Arch Biochem Biophys 1994;311:1–12.
7. Sah RL, Kim YJ, Doong JY, Grodzinsky AJ, Plaas AH, Sandy JD. Biosynthetic response of cartilage explants to dynamic compression. J Orthop Res 1989;7:619–636.
8. Gray ML, Pizzanelli AM, Lee RC, Grodzinsky AJ, Swann DA. Kinetics of the chondrocyte biosynthetic response to compressive load and release. Biochim Biophys Acta 1989;991:415–425.

9. Ando J, Komatsuda T, Ishokawa C, Kamiya A. Fluid shear stress enhanced DNA synthesis in cultured endothelial cells during repair of mechanical denudation. Biorheology 1990;27:675–684.
10. Ookawa K, Sato M, Ohshima N. Changes in microstructure of cultured porcine aortic endothelial cells in the early stage after applying fluid-imposed shear stress. J Biomech 1992;25: 1321–1328.
11. Masuda M, Fujiwara K. Morphological responses of single endothelial cells exposed to physiological levels of fluid shear stress. Med Biol Eng Comput 1991;29(Suppl):268.
12. Frangos JA, Eskin SG, McIntire LV, Ives CL. Flow effects on prostacyclin production by cultured human endothelial cells. Science 1985;227:1477–1479.
13. Rubanyi GM, Romero JC, Vanhoutte PM. Flow-induced release of endothelium-derived relaxing factor. Am J Physiol 1986;250:H1145–H1149.
14. Cooke JP, Stamler J, Andon N, Davies PF, McKinley G, Loscalzo J. Flow stimulates endothelial cells to release a nitrovasodilator that is potentiated by reduced thiol. Am J Physiol 1990; 259:H804–H812.
15. Korenaga K, Ando J, Tsuboi H, Yang Y, Toyo-oka I, Kamiya A. Laminar flow stimulates ATP- and shear stress-dependent nitric oxide production in cultured bovine endothelial cells. Biochem Biophys Res Commun 1994;198:213–219.
16. Diamond SL, Eskin SG, McIntire LV. Fluid flow stimulates tissue plasminogen activator secretion by cultured human endothelial cells. Science 1989;243:1483–1485.
17. Ohtsuka A, Ando J, Korenaga R, Kamiya A, Toyama-sorimachi N, Miyasaka M. The effects of flow on the expression of vascular adhesion molecule-1 by culture mouse endothelial cells. Biophys Biochem Res Commun 1993;193:303–310.
18. Resnik N, Collins T, Atkinson W, Banthron DT, Dewey CF Jr. Platelet-derived growth factor B chain promoter contains a cis-acting fluid shear stress-responsive element. Proc Natl Acad Sci USA 1993;90(10):4591–4595.
19. Wang N, Bulter J, Ingber DE. Mechanotransduction across the cell surface and through the cytoskeleton. Science 1993;260:1124–1127.
20. Fung YC. Biomechanics: Mechanical Properties of Living Tissues. New York: Springer-Verlag, 1988.
21. Ingber DE. Integrins as mechanochemical transducers. Curr Opin Cell Biol 1991;3:841–848.
22. Ingber DE, Folkman J. Mechanochemical switching between growth and differentiation during fibroblast growth factor-stimulated angiogenesis in vitro: role of extracellular matrix. J Cell Biol 1989;109:317–330.
23. Sims J, Karp S, Ingber DE. Altering the cellular mechanical force balance results in integrated changes in cell, cytoskeletal, and nuclear shape. J Cell Sci 1992;103:1215–1222.
24. Cemerikic D, Sackin H. Substrate activation of mechanosensitive, whole-cell currents in renal proximal tubule. Am J Physiol 1993;264:F697–F714.
25. Franco A, Lansman JB. Calcium entry through stretch-inactivated ion channels in mdx myotubes. Nature 1990;344:670–673.
26. Gustin MC, Sachs F, Sigurdson W, Ruknudin A, Bowman C. Single-channel mechanosensitive currents. Science 1991;253:800–802.
27. Morris CE, Sigurdson WJ. Stretch-inactivated ion channels coexist with stretch-activated ions channels. Science 1989;243:807–809.
28. Sachs F, Lecar H. Stochastic models for mechanical transduction. Biophys J 1991;59:1143–1145.
29. Sackin H. A stretch-activated potassium channel sensitive to cell volume. Proc Natl Acad Sci USA 1989;86:1731–1735.
30. Sokabe M, Sachs F, Jing Z. Quantitative video microscopy of patch-clamped membranes: stress, strain, capacitance, and stretch activation. Biophys J 1991;59:722–728.
31. Sukharev SI, Blount P, Martinac B, Blattner FR, Kung C. A large-conductance mechanosensitive channel in *E. coli* encoded by mscL alone. Nature 1994;368:265–268.

32. Sadoshima J, Izumo S. Mechanical stretch rapidly activates multiple signal transduction pathways in cardiac myocytes: potential involvement of an autocrine/paracrine mechanism. EMBO J 1993;12:1681–1692.
33. Glogauer M, Ferrier J, McCulloch AG. Magnetic fields applied to collagen-coated ferric oxide beads induce stretch-activated Ca flux in fibroblasts. Am J Physiol 1995;269:C1093–C1104.
34. Oike M, Droogmans G, Nilius B. Mechanosensitive Ca transients in endothelial cells from human umbilical vein. Proc Natl Acad Sci USA 1994;91:2940–2944.
35. Demer LL, Wortham CM, Dirksen ER, Sanderson MJ. Mechanical stimulation induces intercellular calcium signaling in bovine aortic endothelial cells. Am J Physiol 1993;264: H2094–H2102.
36. Arora PD, Bibby KJ, McCulloch AG. Slow oscillations of free intracellular calcium ion concentration in human fibroblasts responding to mechanical stretch. J Cell Physiol 1994;161: 187–200.
37. Xia SL, Ferrier J. Calcium signal induced by mechanical perturbation of osteoclasts. J Cell Physiol 1995;163:493–501.
38. McDonald F, Somasundaram B, McCann TJ, Mason WT, Meikle MC. Calcium waves in fluid flow stimulated osteoblasts are G protein mediated. Arch Biochem Biophys 1996;326:31–38.

Tissue Engineering for Therapeutic Use 1.
Y. Ikada and Y. Yamaoka, editors.

Reconstitution of collagen gel-based transgenic skin for treating human disease

Katsutoshi Yoshizato
Yoshizato MorphoMatrix Project, ERATO, JST, Hiroshima, and Department of Biological Science, Faculty of Science, Hiroshima University, Hiroshima, Japan

Keywords: artificial skin, collagen, collagen gel contraction, diabetes mellitus, fibroblasts, fibronectin, insulin, lectin, transgene.

The skin equivalent devised by Bell et al. was the first successful example among trials to reconstitute tissue-or organ-like structures in vitro utilizing cells and extracellular matrix substances as component materials which are morphologically and functionally equivalent to in vivo counterparts [1]. The skin equivalent was produced by culturing keratinocytes on fibroblast-populated collagen gels. This reconstitution is based upon the ability of self-assembly of epithelial cells and the contractile force of mesenchymal cells, the latter being important for inducing the tissue-like structures associated with collagen morphogenesis [2]. These two inherent characteristics of cells are required for preparing not only the artificial skin, but also other artificial organs and tissues. This review presents a brief summary of the current understanding of the mechanism of collagen morphogenesis. In addition, a trial to develop a transgenic bioskin is introduced which can eventually be used for treating diabetic patients.

The collagen morphogenesis during the formation of the dermis equivalent

The dermis equivalent is prepared by culturing dermal fibroblasts in three-dimensional lattices of collagen fibrils (collagen gels). At the beginning of the culture, collagen gels are rich in water, and fibroblasts do not complete their attachments to the fibrils, being suspended in gels and round in shape. The fibroblasts gradually attach to the fibrils and extend well on them, taking on spindle-like appearances and protruding many filopodia. Concomitantly with these changes at the cell level, the density of collagen fibrils which is originally even everywhere in the gel, greatly increases around the cells while that in intercellular spaces decreases, indicating that fibroblasts pull the fibrils toward themselves. As a result, the gels contract and greatly reduce their volume, expelling water. The collagen gels are originally fragile, cannot be manipulated by a pair of forceps, and

Address for correspondence: Katsutoshi Yoshizato, Developmental Biology Laboratory, Department of Biological Science, Faculty of Science, Hiroshima University, Higashihiroshima, Hiroshima 739-8526, Japan.

are easily disrupted upon receiving physical disturbances. As the contraction proceeds, they acquire physical strength and finally show characteristics of the dermal equivalent: they suppress cell divisions and enter the Go phase of the cell division cycle [3]; their activity of the protein biosynthesis is quite low [4,5]; collagen fibrils around fibroblasts are arranged parallel to the surface [6]; and morphological characteristics of fibroblasts resemble those in the dermis [7].

The above-described facts about the dermal equivalent convinced us that the reconstituted dermis is a useful experimental model to investigate the mechanism of recognition and binding between fibroblasts and collagen fibrils which eventually lead to the formation of the dermal structure.

Seeking the molecules involved in recognition and binding between fibroblasts and collagen fibrils during the reconstitution of the dermis was done utilizing lectins as probes. We investigated effects of various lectins on the contraction of collagen gel to check the possibility that cell surface glycoproteins of fibroblasts played a role as receptors for collagen fibrils [8]. We tested altogether 12 lectins and selected the following three lectins among them as useful probes to seek for such cell surface glycoproteins: PHA (phaseolus vulgaris agglutinin), WGA (Wheat germ agglutinin), and PNA (peanut agglutinin). PHA suppressed the gel contraction, and inhibited the spreading of fibroblasts on collagen fibrils, but neither on plastic nor gelatin. WGA also suppressed the gel contraction, but did not show such a specificity manifested by PHA for the inhibition of cell spreading. This lectin inhibited the spreading on collagen fibrils, gelatin, and plastic. PNA effected neither the gel contraction, nor the spreading on either of above-mentioned substrates. Thus, PHA was considered to be a useful probe to identify a specific cell surface glucoprotein which mediates binding between collagen fibrils and fibroblasts. We tried to identify a cell surface protein that reacts with PHA, but neither with WGA nor PNA. Proteins bound to each of the three lectins were separated by polyacrylamide gel electrophoresis in the presence of sodium dodecyl sulfate (SDS-PAGE) and were compared among the three. As a result, 250 K protein was selected as a candidate which conforms to the above-described unique mediator between fibroblasts and collagen fibrils. Specific polyclonal antisera were raised against this 250 K protein [9]. The antibody was shown to inhibit the collagen gel contraction in a dose-dependent manner, and immunocytohistochemically detected proteins distributed on the surface of fibroblasts. A cDNA library was constructed from mRNAs of human fibroblasts and expressed by *Escherichia coli*. Utilizing the above antisera, positive cDNA clones were isolated and subcloned for the sequence analysis. The sequence obtained revealed that the cDNAs encode parts of fibronectin (FN). The western blot analysis confirmed that antibodies raised against authentic human plasma fibronectin react with 250 K protein and the antisera against this protein react with authentic human fibronectin. At present, we propose that the 250 K protein is a polymerized form of cellular fibronectin and plays an essential role in mediating the fibroblast to collagen fibril binding which is responsible for the ensuing contraction of collagen gels.

Seeking the molecules involved in the recognition and binding between fibroblasts and collagen fibrils during the reconstitution of the dermis utilizing a specific monoclonal antibody

As another approach to identifying the cell surface molecules participating in the binding between fibroblasts and collagen fibrils as an initial process of the collagen gel contraction, we tried to clone hybridomas producing the monoclonal antibodies which inhibit the contraction of fibroblast-populated collagen gel. Mice were immunized with human fibroblasts and hybridomas were obtained by fusing the immune splenocytes with myeloma cells. Among 98 hybridoma cell lines which produced the antibodies reacting on fibroblast cell surfaces, a clone named A3A5 secreted IgG which inhibited the initial phase of gel contraction [10]. Western blotting showed that the antigen against mAb A3A5 is fibronectin.

There were two completely different and independent approaches to identify the molecules which play a role in the collagen morphogenesis that resulted in the finding of fibronectin, strongly suggesting its pivotal function in the process of dermal reconstitution. The action of A3A5 antibodies was as collagen fibril-specific as PHA: mAb A3A5 inhibited the spreading of fibroblasts on collagen fibrils, but neither on plastic nor gelatin; immunocytohistochemistry with mAb A3A5 stained the surfaces of fibroblasts cultured on gelatin or plastic but not on collagen fibrils.

As is well-known, fibronectin has several functional domains on the molecule responsible for varied biochemical reactions. The next question was knowing which domain was involved in the binding between fibroblasts and collagen fibrils. We performed analytical experiments to answer the question, and aimed at identify the A3A5 epitope to understand the mode of binding among collagen fibrils, fibronectin, and fibroblasts in the process of gel contraction. Fragments containing various parts of fibronectin were prepared by digesting it with pepsin or by expressing recombinant cDNA encoding specific regions of fibronectin in *E. coli*. These fragments were subjected to the western blotting using mAb A3A5 as a probe. As a result the epitope could be located in a region ranging from III_4 to III_6 modules which was different from the gelatin binding domain. Testing the reactivity of fragments within this region produced by the polymerase chains reaction with mAb A3A5, the epitope was further shortened to a 122-amino acid sequence in the III_4 repeat through the amino half of III_5 repeat [11]. Currently available data lead us to suggest that this specific short segment of FN is a novel domain and plays a biologically important role in the complex interactions between cells and collagen fibrils taking place on the cell surface which lead to the formation of the dermis.

The dermal equivalent which produces proinsulin and can be used to treat diabetic animals

In the previous sections, studies to reveal the basic mechanism underlying the process of dermal reconstitution were described. The cell-incorporated hybrid artificial skin appears to be one of the ideal transgene carriers for the gene therapy, because the transplantation and removal (when necessary) of the skin is easily and safely performed in a common operation room and the expression of transgene could be exogenously regulated by treating the transplant with an appropriate stimulator or suppressor. In this section, a novel trial is introduced to utilize the artificial skin composed of epidermal cells and the dermal equivalent as a transgene carrier for the somatic gene therapy. Our current target is to devise a proinsulin-secreting hybrid artificial skin which can be used to treat diabetic animals.

Skin fibroblasts (RSFs) of Fisher 344 rats were transfected with plasmid vectors containing human proinsulin cDNAs (pBMG-Neo-Ins) which episomally propagate in the host cells and actively express cDNAs under the control of metallothionein promoters [12]. The cells incorporating the pBMg-Neo-Ins were selected as surviving cells in medium containing G418, a neomycin analogue, and designated as RSF 308 cells. Fibroblasts transfected with insert-free vectors were also similarly prepared (RSF1518) and used as control cells together with parent cells (RSFs). RSF 308 cells produced immunoreactive insulin (IRI) at 319 ± 22.8 μIU/10^6 cells/h (mean ± SD). Immunocytohistochemistry with anti-human proinsulin antibody showed that about 40% of RSF 308 cells expressed the transgene.

A problem occurred which hampered the preparation of a practically usable hybrid bioskin. The fibroblasts transfected with the vectors greatly reduced their contractile activity and the dermal equivalents prepared with RFS 308 cells or RSF 1518 cells were very fragile and not suitable for the transplantation experiment. This problem was overcome by incorporating parent fibroblasts in addition to RSF 308 cells. The artificial dermis in which 25% of fibroblasts were parent cells showed a contraction comparable to that composed of parent fibroblasts alone.

A total of three types of hybrid skin (HS) were reconstituted in 35-mm plastic dishes by incorporating 1.5×10^6 RSF 308 cells and 0.5×10^6 parent fibroblasts into 2 ml of collagen gels and then placing 10^6 rat keratinocytes on the gels: 1) HS prepared with nontransfected fibroblasts (HS/RSF); 2) with proinsulin producing RSF 308 cells (HS/308); and 3) with fibroblasts transfected with pBMG-Neo (HS/1518). Immunohistochemical studies with anti-proinsulin monoclonal antibodies showed that the rate of proinsulin-positive RSF 308 cells in HS/308 was decreased to approximately 20% of that in monolayer cultures. Most of the positive cells were distributed in the edge of the skin. The average IRI level in culture medium of HS/308 was 178 μIU/10^6 cells/h.

Nude mice were made diabetic by treating them with 130 mg/kg/day of strepozotocin for 2 days. Animals with glucose levels ranging from 200 to 350 mg/dl were used as diabetic animals. Transplantation of HS/308 to their back significantly improved the glucose level, reducing it to 169 ± 99.4 mg/dl at 28 days post-transplantation. In contrast, those animals with HS/RSF and HS/1518 showed

516 ± 64.8 mg/dl and 505 ± 75.8 mg/dl, respectively. Some mice with HS/308 were in hypoglycemia around 30 days. The average plasma IRI level of animals with HS/308 was 2044 ± 594.6 μIU/ml at 4–5 weeks, while IRI of animals with HS/RSF and HS/1518 was undetectable.

During transplantation experiments, it was noticed that all transplants except HS/RSF developed tumors at 3–5 weeks. Growth of tumors in HS/1518 was much less than that in HS/308. The transplants were subjected to histological inspections. The tumor-forming cells were located in the dermis of HS, but not in its epidermis, and no metastasis was observed. Immunocytochemistry showed that proinsulin-producing cells were seen only in the dermis of HS/308, and were widely distributed with different populations in the dermis.

The HS developed in the present study lacks the glucose-level sensing-system by which the extent of proinsulin secretion can be regulated and the level of plasma glucose can be normally maintained. At present, it is not an easy task to incorporate such a regulation system into fibroblasts.

Plasma IRI levels of animals with HS/308 were extremely high at 4–5 weeks. The substance determined as IRI should be proinsulin, because fibroblasts have no appropriate enzymes for processing proinsulin to insulin. The activity of proinsulin to lower the glucose level is approximately 1/20 of insulin [13,14]. The decreased level of blood glucose found in diabetic nude mice with HS/308 can be explained by the high levels of IRI. It is important to introduce an effective way of processing proinsulin to insulin in the HS system, because proinsulin shows some side effects such as hypertension and the size of HSs required can be greatly reduced. Proinsulin is processed in vivo by enzymes, PC2 and PC3, which are active in pancreatic β-cells. A ubiquitous processing enzyme, furin, also can be used as a processing enzyme for genetically engineered proinsulins which have furin consensus cleavage sequences [15,16]. Thus the furin-sensitive proinsulin is one of the possible ways to process proinsulin to insulin in fibroblasts.

pBMG-Neo plasmid vectors which had been derived from bovine papillomavirus (BPV) DNA [17] were used to introduce human proinsulin cDNAs into fibroblasts [18]. This vector might be useful for a sustained transgene expression because it can propagate with the proliferation of host cells. Moreover, the vector does not integrate into chromosomes of host cells, and might not affect the expression of their genes. Cultures of fibroblasts with pBMG-Neo-Ins produce a high level of proinsulin even after repeated passage. However, the HS reconstituted with these cells developed tumors at the transplantation sites. Other investigators also reported that 69% of subgenomic fragments of BPV viral DNA in the pBMG-Neo vector is responsible for tumorigenesis when rodent cells are transformed in culture by the vector [19–21]. Tumorigenesis was seen in the dermis of the transplanted HS with RSF308 as well as RSF1518. The tumorigenicity should be a problem in clinical application. We are attempting to prepare replicative vectors in fibroblasts which do not show tumorigenesis when transplanted to animals. The development of HS with the ability to process proinsulin to insulin, and those without tumor formation after the transplantation are current targets for the intensive investigations.

In conclusion, the present study demonstrated the usefulness of the proinsulin-producing hybrid-type artificial skin to reduce the high glucose levels of diabetic animals. To our knowledge, this is the first experimental demonstration showing that the gene-manipulated hybrid-type artificial skin is a useful gene carrier for gene therapy. Although several problems remain to be solved to establish clinically applicable insulin-releasing HSs, the present study proposed an idea to develop a novel method of gene therapy for type I diabetes mellitus patients. The procedure described in this study might be applied for other diseases as well.

References

1. Bell E, Ehrlich HP, Buttle DJ, Nakatsuji T. Living tissue formed in vitro and accepted as skin-equivalent tissue of full thickness. Science 1981;211:1052–1054.
2. Bell E, Ivarsson B, Merrill C. Production of a tissue-like structure by contraction of collagen lattices by human fibroblasts of different proliferative potential in vitro. Proc Natl Acad Sci 1979; 76:1274–1278.
3. Kono T, Tanii T, Furukawa M, Mizuno N, Kitajima J, Ishii M, Hamada T, Yoshizato K. Cell cycle analysis of human dermal fibroblasts cultured on or in hydrated type I collagen lattices. Dermatol Res 1990;282:258–262.
4. Lambert CA, Martens H, Nusgens BV, Lapier ChM. Regulation of collagenase and Col I Al genes, but not β-actin gene in fibroblasts populated collagen gel is mediated via tyrosine kinase and phospholipase C. J Invest Dermatol 1993;100:436.
5. Lambert CA, Soudant EP, Nusgens BV, Lapiere ChM. Pre-transnational regulation of extracellular matrix macromolecules and collagenase expression in fibroblasts by mechanical forces. Lab Invest 1992;66:444–451.
6. Yoshizato K, Taira T, Yamamoto N, Sasaki K. Remodeling of collagen: an in vitro model of connective tissue. Biomed Res 1985;6:287–296.
7. Grinnel F, Lamke CR. Reorganization of hydrated collagen lattices by human skin fibroblasts. J Cell Sci 1984;66:51–63.
8. Asaga H, Yoshizato K. Recognition of collagen by fibroblasts through cell surface glycoproteins reactive with Phaseolus vulgaris agglutinin. J Cell Sci 1992;101:625–633.
9. Noro N, Ono M, Miyazaki K, Obara M, Yoshizato K. (Manuscript in preparation 1997).
10. Asaga H, Kikuti S, Yoshizato K. Collagen gel contraction by fibroblasts requires cellular fibronectin but not plasma fibronectin. Exp Cell Res 1991;193:167–174.
11. Obara M, Yoshizato K. A novel domain of fibronectin revealed by epitope mapping of a monoclonal antibody which inhibited fibroblasts-mediated collagen gel contraction. FEBS Lett 1997;412:48–52.
12. Yoshida S, Yamasaki C, Yoshizato K. Proinsulin-secreting hybrid artificial skin which can be used to treat diabetic animals. Tissue Eng 1997;3:243–255.
13. Bergenstal RM, Cohen R.M, Lever E, Polonsky K, Jaspan J, Blix PM, Revers R, Olefsky JM, Kolterman O, Steiner K, Cherrington A, Frank B, Galloway J, Rubenstein A H. The metabolic effects of biosynthetic human proinsulin in individuals with type I diabetes. J Clin Endocrinol Metab 1984;58:973–979.
14. Davis SN, Ansiferov M, Hetherington C, Brown M, Branch WJ, Hales CN, Orskov H, Alberti KGMM. The effects of human proinsulin on glucose turnover and intermediary metabolism in insulin-dependent-diabetes mellitus. J Clin Endocrinol Metab 1992;75:1282–1288.
15. Yanagita M, Hoshino H, Nakayama K, Takeuchi T. Processing of mutated proinsulin with tetra-

basic cleavage sites to mature insulin reflects the expression of furin in nonendocrine cell lines. Endocrinology 1993;133:639–644.

16. Groskreutz DJ, Sliwkowski MX, Gorman CM. Genetically engineered proinsulin constitutively processed and secreted as mature, active insulin. J Biol Sci 1994;269:6241–6245.
17. Karasuyama H, Melchers F. Establishment of mouse cell line which constitutively secrete large quantities of interleukin 2, 3, 4, and 5 using modified cDNA expression vectors. Eur J Immunol 1988;18:97–104.
18. Kawakami Y, Yamaoka T, Hirochica R, Yamashita K, Itakura M, Nakauchi H. Somatic gene therapy for diabetes with an immunological safety system for complete removal of transplanted cells. Diabetes 1992;41:956–961.
19. Lowy DR, Dvoretzky J, Shaber R, Law MF, Engel L, Howley PM. In vitro tumorigenic transformation by a defined subgenomic fragment of bovine papilloma virus DNA. Nature 1980; 287:72–74.
20. Schiller JT, Vass WC, Lowy DR. Identification of a second transforming region in bovine papillomavirus DNA. Proc Natl Acad Sci USA 1984;81:7880–7884.
21. Schiller JT, Vass WC, Vousden KH, Lowy DR. E5 open reading frame of bovine papillomavirus type I encodes a transforming gene. J Virol 1986;57:1–6.

Index of authors

Keyword index

Announcing

Tissue Engineering for Therapeutic Use 2

Proceedings of the Second International Symposium of Tissue Engineering for Therapeutic Use, Tokyo, 30-31 October 1997

Editor
Yoshito Ikada
Research Center for Biomedical Engineering
Kyoto University, Kyoto, Japan

Co-editor
Shoji Enomoto
Second Department of Oral and Maxillofacial Surgery
Faculty of Dentistry, Tokyo Medical and Dental University, Tokyo, Japan

Contents

© 1998 152 pages
ISBN 0 444 50076 6 - Hardbound
Price: NLG 132.50/US$76
Excerpta Medica International Congress Series 1170

An Imprint of
Elsevier Science